Recovery of Values from Low-Grade and Complex Minerals

Scrivener Publishing
100 Cummings Center, Suite 541J
Beverly, MA 01915-6106

Publishers at Scrivener
Martin Scrivener (martin@scrivenerpublishing.com)
Phillip Carmical (pcarmical@scrivenerpublishing.com)

Recovery of Values from Low-Grade and Complex Minerals

Development of Sustainable Processes

Edited by
Elvis Fosso-Kankeu
Department of Metallurgy, University of Johannesburg, South Africa
Bhekie B. Mamba
College of Science, Engineering and Technology, University of South Africa, Pretoria, South Africa
and
Antoine F. Mulaba-Bafubiandi
Emeritus Professor at the University of Johannesburg, South Africa

Scrivener Publishing

WILEY

This edition first published 2024 by John Wiley & Sons, Inc., 111 River Street, Hoboken, NJ 07030, USA and Scrivener Publishing LLC, 100 Cummings Center, Suite 541J, Beverly, MA 01915, USA

For more information about Scrivener publications please visit www.scrivenerpublishing.com.

Wiley Global Headquarters
111 River Street, Hoboken, NJ 07030, USA

For details of our global editorial offices, customer services, and more information about Wiley products visit us at www.wiley.com.

Library of Congress Cataloging-in-Publication Data

ISBN 978-1-119-89641-8

Cover image: Pixabay.Com
Cover design by Russell Richardson

Set in size of 11pt and Minion Pro by Manila Typesetting Company, Makati, Philippines

Printed in the USA

10 9 8 7 6 5 4 3 2 1

Contents

Preface

The mineral industry is undergoing significant cultural, organizational, and technological transformations to address some of the major limitations and challenges related to the environmental and productivity domains. As far as productivity is concerned, the decrease of high-grade ores has been one of the stumbling blocks toward the achievement of maximum recovery of metals while, on the other hand, the complexity of minerals therein makes it difficult to profitably extract metals using only conventional methods. Recent developments in comminution, froth flotation, and leaching processes have focused mainly on the improvement of technologies to minimize the impact of gangue materials on the recovery of values and therefore increase the grade of minerals or the recovery rate of metals, as well as minimizing the environmental impact of extractive metallurgy activities. In fact, a sustainable recovery of values from minerals implies a cost-effective process, but also a cleaner and responsible production that suggests an ethical and preventative strategy that reduces the risk of environmental pollution.

The refractoriness of metal sources including ore, flotation concentrates, mill tailings and others that could affect mineral/metal processing could be due either to the fine dissemination of the metal in minerals or gangue matrixes that can react with metal host minerals, increasing the consumption of lixiviant and decreasing the rate of leaching reactions. On the other hand, the formation of intermediate phases or species may occur during the leaching process and form passivation layers, or consume important lixiviation reagents, and both will negatively affect the extraction of metal. There are also challenges associated with the greater mineralogical complexity of low-grade minerals from which metals can be extracted.

After the comminution, froth flotation constitutes an important segment of the separation process in the mineral processing industry and was introduced several decades ago to effectively separate gangue materials from valuable minerals. Despite being a matured process, the complexity of minerals can make froth flotation suffer from a relative inefficiency that could be attributed to poor insight regarding the physical and chemical phenomena that underpin the relationship between surface wettability and floatability. For instance, with the decrease of high-grade minerals coupled with the increased complexity of mineral composition and dissemination, the mineral industry has devoted its attention to low-grade and complex ores with high clay contents. However, to liberate values from such ores requires that they are ground to very fine sizes, which also contributes to the liberation of large quantities of fine gangue that are likely to cause problems related to slime coating, rheology, and entrainment during the flotation process. These issues mainly result in overconsumption of reagents, reduction of direct contact between the targeted mineral and collectors, and the reduction of the grade of flotation concentrations that contribute to minimizing the efficiency of the flotation process.

For all these cases, a pre-treatment of the mineral is needed to reduce the amount of the gangue upfront or adapt an adequate technique that will effectively control the formation of unwanted transition phases that can negatively affect the recovery of values.

The technological imperative will be therefore to improve the efficiency and reliability of the processes considered for metal extraction from the source. This technology should additionally be cost-effective and ensure the minimal negative impact on the environment. The various challenges encountered in the industry regarding the processing of complex minerals, research questions raised from unsolved technological mysteries, as well as the research illustrations presented are the various research benefits the readers will find in this book.

This book presents eight specialized chapters that focus on the exploration of the complexity of minerals that are likely to negatively influence the recovery of values, as well as the development of adequate technologies capable of improving the process of mineral concentration and/or metal recovery from complex minerals in a sustainable manner.

This book reviews the various physicochemical properties of minerals that are likely to pose a challenge during the attempt to recover values using conventional methods. It also elaborates on the recent technological development that has been considered by researchers to improve the recovery of metals from gangue-dominated minerals while ensuring cost-effectiveness and minimal adverse environmental impact.

The editors and the publisher are grateful to the reviewers who have contributed to improving the quality of the book through their constructive comments. The editors also thank the publisher for including this book in their series.

This book will be of interest to academic researchers in the fields of mineral processing, hydrometallurgy, geochemistry, environment, chemistry, engineering, and professionals, including mining plant operators, environmental managers in the industries, government regulatory officers, and environmentalists.

Elvis Fosso-Kankeu
Bhekie B. Mamba
Antoine F. Mulaba-Bafubiandi
January 2024

1

Optimization of the Mechanical Comminution – The Crushing Stage

Ngonidzashe Chimwani

Department of Mining Engineering, University of South Africa (UNISA), Florida Campus, Johannesburg, South Africa

Abstract

The mineral processing industry is faced with various challenges that hinder the efficient extraction as well as the marketing of metals. Among those challenges are the depletion of high-grade ores, rise in energy costs, and reduction of metal prices. In view of that, the need to improve efficiency becomes eminent in the mining and mineral processing industry. Improving efficiency can be achieved through carefully reviewing all the mining and mineral processing stages to identify technical innovations that enable the establishment of high-efficiency and low-energy consumption processes. This chapter thus reviews different crushers and the parameters that influence their potential to process low-grade ores efficiently. The review unveiled that downstream processes-oriented research should be preferred to those that focus on the optimization of a particular unit to improve efficiency. Other initiatives, such as managing the demand side of power and optimizing the idling time of the crushers, have shown great potential to reduce the energy consumption of impact crushers. This is because the energy expenditure of a gyratory crusher during idling is 30% of the full load power consumption, and jaw crushers not only lose a significant portion of energy into noise and heat, but have no-load power ranging from 40% to 50%.

***Keywords*:** Comminution, jaw crusher, gyratory crusher, cone crusher, impact crusher, HPGR crusher, low-grade ore, optimization

Email: ngodzazw@gmail.com

Elvis Fosso-Kankeu, Bhekie B. Mamba and Antoine F. Mulaba-Bafubiandi (eds.) Recovery of Values from Low-Grade and Complex Minerals: Development of Sustainable Processes, (1–40) © 2024 Scrivener Publishing LLC

1.1 Introduction

Comminution involves processes that reduce rocks from the infinite size of the half-space to sizes that allow valuable minerals to be separated from the gangue [1] using different separation methods. Rock blasting is the first process of comminution followed by crushing and then grinding, and the efficiency of each process is dependent on the preceding process. The blasting being on the utmost upstream has to provide not only the desired particle sizes for the crushers and mills but particle grains with reduced internal resistance owing to the existence of macro and microfissures within the particle. Although the presence of flaws within the blasted particles enhances productivity and lowers the grinding energy during mechanical comminution generally, they cause a reduction in productivity in autogenous mills that use the blasted particles as the grinding media. Thus, the control of the blasting stage should be guided by the requirements of the downstream process.

Requirements of the downstream processes are further complicated by the challenges exacerbated by the processing of low-grade ores resulting from the depletion of high-grade ores. Low-grade ores contain small grain-sized minerals dispersed through the rock, which require high energy as well as the knowledge of the ore texture to free the minerals from the host, and worse still, high fine losses are unavoidable in the process. The high energy requirements of low-grade ore processing add to the headache that the mineral processing industry has had since its inception. The comminution stage has, over the years, been widely known to be characterized by low efficiency [2] and high energy consumption, among other things. The continuous rise in energy costs and the increased concerns over global warming have sparked efforts to minimize CO_2 emissions, albeit with costs. Thus, it is imperative to be proactive than reactive, that is, to address the problems at the design and process stages.

To that end, several efforts have been made in that direction, at the crushing and milling stages including the development of comminution tests. The tests conducted are the drop weight tests, which provide the abrasion properties of rocks; the bond crushability work index, which determines the energy requirements of a crusher; the bond abrasion tests, which estimate mill/crusher liners; and the point load tests, which measure crushability by correlating the mechanical strength of the rock and the crusher reduction ratio. All these tests play an important role in the simulation and design of crushers and mills. The information drawn from these tests is used by the designers and manufacturers of comminution equipment to determine the load and non-load power requirements of each crusher/mill. From the wealth of research also emerged that the random application of forces inside the crushing and grinding machines between particles themselves or particles and the machine parts is fundamental to the disruptive nature of the comminution process [3]. These random forces make it impossible to achieve 100% efficiency in the comminution machines. Variations of the interactions of those particles and the products obtained are influenced by the equipment and operational parameters, such as the shaft velocities, equipment geometry and material properties, which ultimately affect the process efficiency in general, and the equipment in particular. Some equipment effect single impact, while others apply multiple impacts until particle sizes are disintegrated and reduced to acceptable values. The challenge, however, is that those processes are not efficient as mentioned in the foregoing, worse still with the processing of low-grade ores. This is especially considering that the biggest fraction of the input energy is dissipated as noise and heat with the other fraction channeled towards disrupting the forces that bind particles constituting the ore [4]. Richardson and Harker [5] compiled a list of the breakdown of the energy expenditure in comminution processes, and in that list, energy was observed to be consumed through the production of elastic deformation of the particles before fracture occurrence, inelastic deformation that leads to size reduction, elastic equipment distortion, friction between particles and between particles and the machine, and in the vibration, heat, and noise in the plant.

The proper definition of energy expenditure has been the subject of research over the years, and the need to deepen the enquiry cannot be overemphasized especially with the current dominance of low-grade ore processing. It is important to mention that the widely accepted theory used to determine the actual amount of energy used in comminution is derived from the inverse proportionality relationship between the mean particle size of the particles and the newly produced surface area, as the milling progresses. This thus means that the determination of surface areas at different stages of milling provides a reliable inquiry into the energy expenditure of the comminution process. In that vein, the Bond work index has been widely used and found to be reliable [6]. From comparing the total ideal fracture surface energy needed to generate new surfaces during the comminution of covalent materials and the standard Bond work index, Tromans [7] observed that crushing and grinding operations have 1% energy efficiency or even less.

Modeling energy consumption of a comminution equipment requires the knowledge of the design and operational parameters of the equipment [6]. The parameters range from that of the feed material to those of the equipment comminuting the materials. Some researchers have suggested altering the parameters of the feed through pre-treatment of the material to be comminuted to reduce Bond index values [3]. Although the pre-treatment increases cracks in the particles, it has the possibility of consuming more energy than the comminution process itself [6]. The most followed procedure, however, is to optimize the comminution process through fine-tuning the equipment parameters in a manner that redirects energy to be spent through vibration, heat, and noise into strain energy, which alters the internal forces of particles, hence creating new surface areas. But optimizing the process undoubtedly requires an understanding of Tromans' [7] detailed account of the energy expenditure. The author observed energy as utilized through the elastic deformation of particles before fracturing occurs, the inelastic deformation that results in the breakage of particles, elastic distortion of the equipment, friction between particles and between particles and the machine, and in noise, heat, and vibration of the equipment. It is also important to note that energy efficiency during comminution, as noted by Tromans [7] depends on impact efficiency, which, in turn, is influenced by the particle loading force, size, and orientation of inherent fissures in the particle with respect to the loading axis.

Over the years, the design and optimization of comminution equipment were achieved through experimental work and important milestones were reached. However, the experimental approach could not show how the particles flow and interact among themselves and with the equipment parts. Researchers thus found computer modeling applicable to unpack the internal dynamics of the particles that cause breakage. Advanced tools that allow particle flows to be simulated and the breakage process to be predicted, such as the discrete element modeling (DEM), were utilized. The tool found use in soil mechanics [8], and then later applied to milling and mineral processing [9]. Its ability to address a wide range of particle-related problems is well-documented in the literature [10–12]. Recently, it was used to simulate solid flows and energy transfer in the vertical impact crusher [13], configure ball mill liners to induce ball segregation in a ball mill [14], explore end liners of a ball mill [14], and simulate an impact crusher [15]. Its extensive usage in modeling particulate systems makes it useful to optimize comminution processes, which reduce particle sizes of low-grade ores.

The population balance model (PBM) was also used widely in the optimization of comminution circuits. The model, which found more use on the milling side than the crusher side to model size reduction kinetics, has been successfully applied to modeling hammer mills [16]. The PBM concept is used in the formulation of grinding equations derived from mathematically relating feed size and product size through applying size-mass balance. It is thus defined by the selection of particles for breakage (selection function) and the breakage distribution of 'children' fragments as a result of broken 'parent' particles (breakage function) [6] commonly known as the specific rate of breakage and the primary breakage distribution functions, respectively. These two kinetic functions form the basis for the modeling of the breakage kinetics in comminution devices when combined with the mass transportation of particles [17]. In this scheme, size-class mass balance calculations are used to predict the mass change of size classes at any given time interval from the milling of a narrow size class with set boundaries. From the crushing side of comminution, Chimwani and Bwalya [18] modeled the kinetics of an impact crusher using the PBM while for milling, some researchers have coupled it with other graphical analysis tools, such as the attainable region (AR) to optimize comminution circuits [19], with insightful information obtained from their findings.

All the efforts done so far to lower energy consumption through the optimization of the mechanical comminution stage are impressive. However, the problem is still very far from being contained since high-grade ore is depleting, resulting in the use of low-grade ore, which further complicates the optimization scheme. Furthermore, achieving 100% efficiency in comminution is something next to impossible, as observed by Legendre [20]. In the subsequent sections, different crushers are reviewed with the idea of bringing to light different efforts put by different researchers to understand their mode of operation in an attempt to achieve optimum operation of those size-reduction machines. This would consequently enable feasible processing of low-grade ore and the required crushing proportion of the comminution process.

1.2 The Role of Crushers

Crushers are at the initial stage of the mechanical comminution in the mineral processing chain. They compress or impact blasted particles in dry form against a rigid surface in a constrained motion path into different sizes and shapes for further downstream processing. The run of mine (ROM) ore is crushed in the succession of the primary, secondary, and tertiary crushers, and the particle size and shape requirements of the final product determine the ultimate choice of each crusher in a comminution circuit [21]. The crushing mechanism that dominates the breakage of particles in the crusher also plays a very crucial role in its selection for a given task. This is because some crushers consist of single-pass events that have been associated with the production of angular particles, while those with retention systems have been linked to the production of more rounded particles [22]. Particles that are angular and irregular, with sharp edges, possibly caused by the intersection of the propagated cracks, also result from the massive fracture of particles as effected by some crushers, while the rounded particles result from the chipping of edges and the erosion of surfaces as effected by other types of crushers. How the particles are fed into the crushers also influences the particle shape of the product, as observed by Durney and Meloy [23].

1.2.1 Types of Crushers and Their Effect

1.2.1.1 Jaw Crusher

A heavy-duty machine classified as a primary crusher, a jaw crusher reduces the particle size of the ROM ore in the order of 1.5-m size down to the product of 0.1 – 0.2 m dimensions by compressive force. It is a widely used comminution process equipment characterized by the V-shaped mouth at the top of the crushing chamber. The crushing chamber consists of a flat fixed rigid jaw and the flat or convex pivoting jaw, with hardened steel faces, which swing and thus effect compression between the swinging and stationary plates. The jaws are set at acute angles to each other, and the product size is determined by the setting at the bottom of the plates. Parameters, such as the strength, feed and product size, cost, and volume of operation, among other operational parameters, determine the optimal operation of the machines [24]. It is important to mention that the type of ores to be processed determines which jaw crusher to select for a given task. The swing jaw pivoting position classifies jaw crushers into Blake, Dodge, and Universal crushers, which in turn determines which types of ores the crusher can process.

The Blake crusher, which can be further classified into the single- and double-toggle jaw crusher, has a lower position fixed swing jaw with a variable discharge and fixed feed areas. In a single-toggle crusher, the swing jaw undergoes an elliptical jaw motion characterized by the swing motion and vertical movement caused by the toggle plate and the rotation of the eccentric angle, respectively. Whereas, in a double toggle as shown in Figure 1.1, the vertical motion of the pitman effects the swing jaw oscillating motion. This makes the single toggle suitable for softer materials, while the double toggle efficiently crushes the tough and abrasive materials. Therefore, double-toggle crushers are commonly used in mineral industries as primary crushers [6].

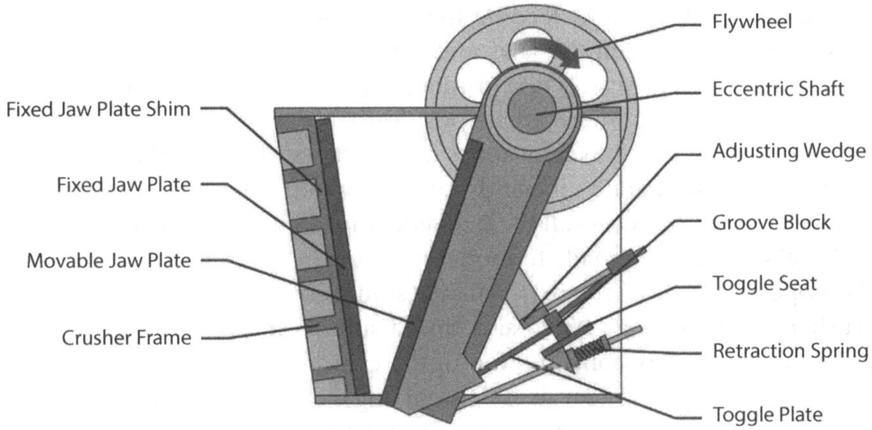

Figure 1.1 Illustration of a double jaw crusher [37].

In that light, all the factors have to be borne in mind when selecting crushers, especially when processing low-grade ores, which are more expensive to process. Important design factors are as follows [6]:

$$\text{Crusher vertical height} \approx 2 \text{ x Gape}$$

$$\text{Jaw width} > 1.3 \text{ x Gape}$$

$$< 3.0 \text{ x Gape}$$

$$\text{Throw} = 0.0502\ (\text{Gape})^{0.85} \tag{1.1}$$

where the crusher gape is in meters.

To size the gape, the following relationship between the gape and the largest particle is used [6]:

$$\text{Largest particle size} = 0.9 \text{ x Gape} \tag{1.2}$$

The swing jaw of the Dodge crusher is fixed at the upper position, and that fixes its discharge area while the feed area is variable. The crusher operates under choking conditions with particles broken smaller and smaller progressively as they approach the bottom end and eventually escape. This makes the crusher unsuitable for handling heavy-duty crushing and hence limited to laboratory use. It is important to note that operating crushers under choking conditions have a bearing on the product particle shape.

In addition to that, choke feeding makes particles break because of compressive forces between the jaws and inter-particle compression with the consequence of a generally fine product [6]. The universal crusher has the swing jaw fixed at an intermediate position.

Jaw crushers, being usually at the initial stages of the mechanical comminution, must operate efficiently so that the efficient operation culture can be transmitted across all the size reduction stages. In that vein, considerable effort has been made to model the crushing machines and optimize their operations [25]. The performance studies include optimizing the jaw crusher in terms of the speed and closed side setting [25, 26], kinematics [27], and the energy consumption of the crusher [3, 28–30]. Some of the researchers enlisted the services of modeling techniques, such as the DEM, to assess the variation of the crusher product with ore body strength [31], response surface method and gene expression programming [32], variable load-based optimal control [29] while others optimized the crusher, in terms of particles using the population balance model [33].

Parameters of jaw crushers that determine their optimal operation are the crusher speed, closed side setting (CSS), feed rate, and gradation. The CSS, determined by the smallest gap between the moving and fixed jaws, is the main parameter for jaw crusher optimization and has implications for the energy consumption and capacity of the jaw crusher. It is also known as the aperture of discharge in the crushing medium [32] that controls the extent of rock crushability of compressive crushers [34]. From the work of Barrios *et al.* [35], CSS was found to be central to the estimation of the product achievement, throughout, compressive forces acting on jaw plates, specific energy consumption and the size reduction ratio. An inverse proportionality relationship occurs between the CSS and the power draw of the crusher, while between the crusher capacity and the CSS, it is the direct proportionality relationship [25]. The jaw speed controls the capacity, as well as the fineness of the product at any given CSS, since it determines the number of compressions a particle gets in its journey between the jaws until it exits the crushing chamber. The distance traveled and time taken while subjected to repeated crushing forces in the crushing chamber were used by Rose and English as the basis for determining the capacity of the jaw crusher [3]. Higher speeds of the jaws reduce the capacity of the crusher since it causes the particle to stay longer in the crushing chamber at the cost of high energy expenditure. Although the result is generally a finer product, the high energy consumed in the process diminishes the returns, especially considering the ever-escalating energy costs and the efforts to reduce the emission of CO_2.

Energy loss through the dissipation of the input energy into noise and heat characterize most comminution process, and thus forms a basic approach on which the modeling of the jaw crushers is based [7]. The processes are optimized numerically with machines that have large computational power, which allows the manipulation of the jaw crusher design parameters to determine efficiency values. Some researchers have also developed an optimal switching control technique that reduces the energy consumption of the jaw crusher, hence the crushing process [36]. This was developed after realizing that compressive crushers such as jaw crushers consume 40% to 50% of rated power when idling without a load, hence filling the energy improvement gaps in literature.

In a study aimed at providing information about the single-stage crushing optimization, Fladvad and Onnela [30] conducted a study that investigated the effect of feed rate and gradation, crusher setting and speed on the crusher operation, and quality of the product. The quality of the aggregates produced was determined by the shape, gradation, and the mechanical strength of the feed particles. From that work emerged that the specific energy consumption during crushing is influenced by the feed gradation, crusher setting, and crusher speed. Interesting to note though is that the authors found that crusher speed was inversely proportional to the specific energy consumption of the crushing process. It was also concluded in this work that products from a specific rock deposit should not be assumed to have constant mechanical strength, and that product quality and crushing efficiency can be controlled in a jaw crusher set-up. Mechanical strength depends on the interaction between crystals, particles, and the cementation material forming it [38].

In a study that focused on the effects of rock strength properties on the performance of a jaw crusher, Olaleye [31] concluded that the stronger the rock, the longer the crushing time required to obtain a given grain-size distribution of particles, hence the higher wear and energy consumption. In another study, Köken and Lawal [32] used the response surface methodology to investigate the choke-feeding intensity of feed sizes ranging from 9.5 to 19 mm during crushing actions. Their findings revealed that the characterized feed size distribution controls the general size reduction, while the amount of the feed in terms of mass affects the energy consumed in the crushing process and the product flakiness. Both the feed size distribution and mass parameters were found to be influential in regulating the choke-feed intensity. The authors acknowledged the importance of the choke-feeding intensity parameter in compressive crushers, hence its important role in the optimization of jaw-crushing operations. They suggested that relatively low choke-feeding intensity results in insufficient throughput to effect initial rock breakage. Köken and Lawal [32] also observed that the choke-feed intensity parameter changes as the physico-chemical and mineralogical features of the rock change, which means the comminution optimization of rocks varies with rock types. Different rocks have different uniaxial compressive strengths, which influence how energy is consumed in jaw crushers [39] and are a function of the degree of rock crushability (DRC) [40].

The assessment of the DRC is based on the size reduction ratio (SRR), size distribution width (Δw), product flakiness index (FI_p), and total energy consumption (E_t), which are presented in the following equations [32].

$$E_t \frac{V \times cos\varnothing}{10^3} \sum_{i=0}^{t} I_i \times t_i \tag{1.3}$$

where I is the current drawn by the crusher (A),
V is the voltage (V),
$cos\varnothing$ is the power factor of the engine propelling the crusher,
t is the crushing time (s)

$$SRR = \frac{F_{80}}{P_{80}} \tag{1.4}$$

where F_{80} is the theoretical square mesh aperture sizes (mm) corresponding to 80% of the cumulative undersize of the feed's PS',

P_{80} is the theoretical square mesh aperture sizes (mm) corresponding to 80% of the cumulative product's PS'.

$$\Delta w = \frac{P_{80} - P_{20}}{P_{50}} \tag{1.5}$$

where P_{80}, P_{50}, and P_{20} are the theoretical square mesh aperture sizes (mm) corresponding to 80%, 50%, and 20% of the cumulative undersize of the product's particle size distribution (PSD), respectively.

The FL_p is determined by flakiness index tests guided by the BS EN 933-3.

The detailed rock-crusher investigation enables the achievement of the optimum crushing outputs influenced by the rock properties, feeding method, and the crusher performance. The behavior of rock aggregates under choke-feeding, determined from the crusher tests, provides reasonable grounds to assume that the efficiency of the crushers is hinged on their inherent breakage mechanism. This makes it imperative for mineral processors to choose the right crusher for a given task, especially in the wake of low-grade ore processing.

Gupta and Yan [6] gave the operating functions and the capacity of the jaw crusher. The capacity determined by the mass or volume of the produced crushed material per unit time of operation is a function of the width and depth of the crushing chamber; open-and-closed side settings; options on feeding methods (intermittent or continuous); and operating characteristics (length of stroke, number of strokes per minute, and the nip angle) [6]. The mathematical relationship is as follows:

$$Q = f\,(w, L, L_{MAX}, L_{MIN}, L_T, n, \theta, K) \tag{1.6}$$

where Q = capacity,
w = width,
L = height (or depth of jaws),
L_{MAX} = set maximum (Open Set),
L_{MIN} = set minimum (Closed Set),
L_T = length of Stroke or Throw,
n = frequency (Revolutions rpm = cycles per unit time),
K = constant related to machine characteristics, and
θ = jaw angle.

The capacity of the jaw is significantly affected by the frequency of the opening and closing of the jaws.

1.2.1.2 Gyratory Crusher

The crusher as presented in Figure 1.2 has a long, conical, hard steel crushing element, suspended from the top, which sweeps out in a conical path within the round fixed crushing chamber as it makes slight circular movements [24]. The size reduction mechanism in the crushers is primarily by the compression of a rock by two pieces of steel. The degree of crushing is determined by the gap between the concave fixed liners mounted on the main frame of the crusher and the hard crushing surface attached to the spindle, both of which are lined with manganese steel surfaces. The central vertical spindle gyrates because it is eccentrically driven at the bottom and that causes the gape to open and close, effecting particle breakage. The design of a gyratory crusher is influenced by the size, distance traveled by

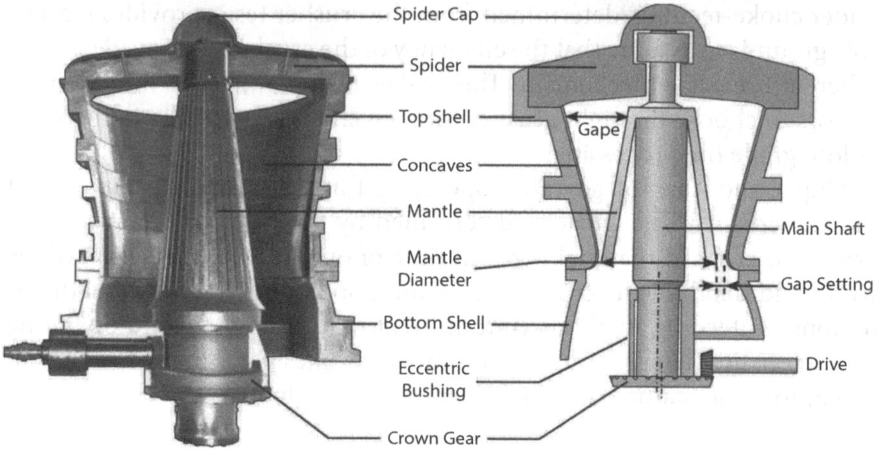

Figure 1.2 Gyratory crusher [37].

particles down the crusher (useful height), set range, speed (revolutions per minute), and power [41].

Gyratory crushers process soft to very hard abrasive material, which is dry to slightly wet but not sticky. They are commonly used to crush heavy mining quarried materials in a mine or ore processing plant with a top size of 1370 mm down to less than 200 mm, resulting in a size reduction ratio of 4 – 7/1. When choke-fed, the crushers are believed to produce material suitable for feeding the semi-autogenous (SAG) mills. Additionally, choke feeding of the gyratory crushers results in the long life and even the wear of the mantle. Their ability to crush on full cycles makes them the right candidates for high-capacity production, as long as the crushing chamber is always full. They are less costly to install, albeit with higher maintenance costs compared to jaw crushers. However, the pros and cons of each crusher offset each other when viewed as a total package.

Over the years, important milestones were reached regarding the optimization of gyratory crushers with some researchers focusing on the characteristics of the crusher [42–44] while others take ore characteristics into consideration [45]. Chen *et al.* [46] combined the DEM analysis based on the bonded-particle model and the response surface methodology to investigate the mantle shaft, concave, and eccentric angles, as well as the mantle speed on the gyratory crusher performance. Their work was motivated by the assertion of Gang *et al.* [43] that it was not enough to optimize the crusher in terms of the mantle shaft speed, eccentric angle, and mantle shaft height without taking the impact of the crushing chamber on the performance into consideration. Also, from their investigation of the effect of liner wear on the crusher performance, Rosario *et al.* [42] linked the performance of the crusher to the crushing chamber and from an optimization scheme, Chen *et al.* (2020) managed to increase the crushing forces by 2.25 times and productivity by 8.4%.

Later, Chen *et al.* [46] noted that the geometric characteristics of the concave (concave curve height and radius) and motion characteristics of the mantle shaft affect the discharge port size of the crusher and consequently, the performance of the gyratory crusher. In that view, they used DEM to investigate the concave curve height, radius, eccentric angle, and mantle shaft speed influence on the crusher chamber performance. The eccentric angle influences the discharge port size and the compression ratio, which in turn affects the mantle shaft crushing force and the granularity of the crusher discharge. Apparently, the mantle shaft speed has a greater impact on the crusher performance than the shape of the crushing chamber, which has very little significance on productivity [45]. From their optimization program, Chen and colleagues observed that the concave curve height and radius strongly influence the crushing force and power density. It was also concluded that mantle shaft speed significantly affects the power density and the granularity of the discharge. Also, for this crusher, it was concluded that productivity is directly proportional to power density. The crushing chamber was at the center of the enquiry owing to its direct interaction with the ore, which in turn has a powerful influence on the quality and cost of subsequent iron purification [42, 43]. Such studies, which are downstream performance-oriented, are the most relevant, especially in the wake of the low-grade ore processing times.

To assess the performance and eventually the productivity of the gyratory crusher, Li *et al.* [45] modeled the crusher based on the kinematic and dynamic characteristics of the material. They used the layered breakage theory and structure of the crushing chamber to determine the trajectories and discharge of the particles. This enabled the authors to compute the upward and downward velocity of the material in the choke layer, hence the productivity. Therefore, it goes without saying that the performance and productivity of the gyratory crusher are also influenced by the ore characteristics such as the fines content of the feed (which should be < 10%), the rock moisture content, feed distribution in the crushing area and the feed bulk density, work index (the rock hardness), and recirculating load (when the crushing circuit is closed). Also observed is that it is important to charge uniform feed when a uniform-size product is desired. Gyratory crushers have exhibited different performances when processing ores with different work indices, thus indicating that low-grade properties of the ore should be well understood in order to fine-tune the crusher to the processing requirements of the ore. It is interesting to note that gyratory crushers consume 0.3 times the full load power during idling, which makes it imperative to avoid unnecessary idling, especially when processing low-grade ore, lest it increases the cost of extracting the metal, hence further dwindling the profit margin.

It is important to mention that gyratory crushers are usually preferred to jaw crushers when the production rate required is more than 900 t/h [41]. The product of a gyratory crusher, which usually ranges from 0.1 to 0.15 m, is normally too big for downstream mineral liberation processes. Thus, secondary, or even tertiary crushers may be needed to form a circuit of crushers with primary, secondary, and tertiary crushers having 3:1 – 10:1, 6:1–8:1, and ≅ 10, respectively. These ratios are used to size the gape and set of the crusher, depending on which stage the crushing is taking place.

To determine key operational parameters such as the gyrating speed, Equation 1.7 is used [41]:

$$v \geq \frac{665(\sin\theta - \mu\cos\theta}{\sqrt{d}} \textit{cycles per minute} \tag{1.7}$$

where θ = inclination of the cone to the horizontal,
μ = coefficient of friction of the material, and
d = size of crusher product (cm).

Among the different methods used to determine the gyratory crusher capacity, the one advocated by Rose and English presented in Equation 1.8 emerged to be the most practical one [41]:

$$Q = W_i D \rho_s \frac{\sqrt{L_{MAX} - L_{MIN}}}{2\sqrt{\frac{R}{R-1}}} \left(L_{MAX} + L_{MIN}\right) K \left(\frac{t}{h}\right) \tag{1.8}$$

where W_i = Bond's work index,
D = diameter of bowl at a given cross-section,
L_{MAX} = maximum distance between the bowl and lower edge of mantle,
L_{MIN} = minimum distance between the bowl and lower edge (closed set),
r = reduction ratio, and
K = statistical factor.

For soft materials, such as coal and coke, $K = 0.5$.

For harder materials, such as quartz and granite, $K = 1$.

For both jaw and gyratory crushers, power is determined using the method developed by Andersen and Napier-Munn [47] from a material balance method proposed by Whiten [48].

1.2.1.3 Impact Crusher

Impact crushers break rocks by impact rather than compression through impact blows at high-speed to the particles that are free-falling. The blows are impacted by the hammers hinged or fixed on to the edges of the rotating discs, both of which are made of cast iron containing chromium carbide or manganese steel. The kinetic energy transferred to the particles when they are impacted, creates internal stress within them and causes them to shatter. The nature of the breakage mechanism in these crushers allows the particles to fragment without leaving residuals in them [49]. This, therefore, makes their products more preferred in the quarry industry, although they are applicable to mineral, cement, coal, and chemical industries. Efforts to understand the crusher were made through mathematical modeling and experimental work [50]. What are still gray areas though are the amount of feed amounts, the rotor speed effect, the mechanism of size reduction in the crusher (determined by the energy distribution and velocity of collisions inside the crushing chamber), the lining design, and the design of the hammer on the impact crusher. Using *t-family* value analysis and the Bond work index approach, investigations by Deniz [50] found the product size reduction to be strongly related to the mean feed size distribution and the Bond work index to vary with material properties.

Impact crushers are classified into horizontal and vertical shaft impactors. The horizontal shaft impactor (HSI), as presented in Figure 1.3, has hammers fixed upon the outer edge of a spinning rotor and has two or more distinct breakage regions [51].

The particles in HSI fall tangentially onto a high-speed rotor and are impacted by blow bars and then liners in the crushing chamber before passing into the next chamber through the clearance between the rotor and liners. The particles undergo the same process until they exit the last chamber. Although the crusher can process hard or soft materials, it has been traditionally limited to the grinding of non-abrasive and soft materials. There is also a multi-shaft HSI that has attached flingers that impact materials dropping into it by gravity. The crusher has three stages, and each stage has two opposite shafts rotating towards each other. The particles within the crushers experience high energy collision from themselves, the flingers, and liners. Different types of ore can be processed in the crusher at high throughputs with high reduction ratios achieved [53].

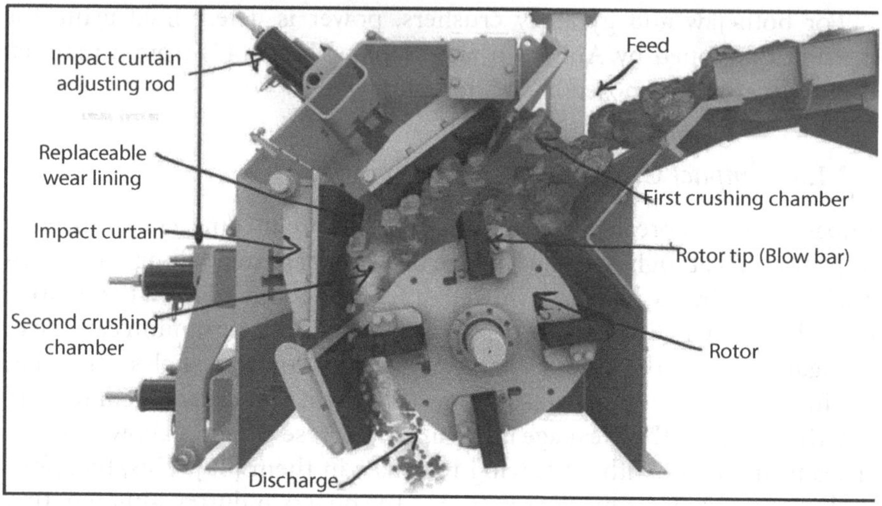

Figure 1.3 Horizontal shaft impactor [52].

The vertical shaft impactor has a single breakage zone and high-speed rotors with tips and chambers that are wear-resistant. In this type of crusher, particles entering the rotor are impacted and flung outward centrifugally into a particle cloud cascading through the crushing chamber. Velocity is utilized as the predominant force to break particles rather than the surface force, owing to its ability to disperse the breaking force through the mass and across the surface of the rock. This enables the crusher to handle rock particles even with much higher abrasiveness and to be widely known as characterized by a higher degree of comminution, high quality of material produced with high-cost effectiveness [54]. There is a type of vertical shaft impactor (VSI) presented in Figure 1.4, which consists of iron blow bars mounted on the rotor contained within the ring of thin anvils. The material impacted by the blow bars is accelerated to the anvils, where they get another impact and shatter before leaving the crushing chamber through the discharge chute. In VSIs, the granular gas produced in their grinding chamber either takes too long to settle or the discharge arrangement is physically restricted, both of which effect restricted discharge and give rise to recycle loads [51].

In an attempt to understand the operation of VSI, Al-Khasawneh [56] used a mathematical- physical model based on the physical and mathematical principles of the motion of a particle to design and test a ring armor for the VSI. The objective function was to enhance the product quality, improve grinding fineness, and ultimately reduce energy consumption.

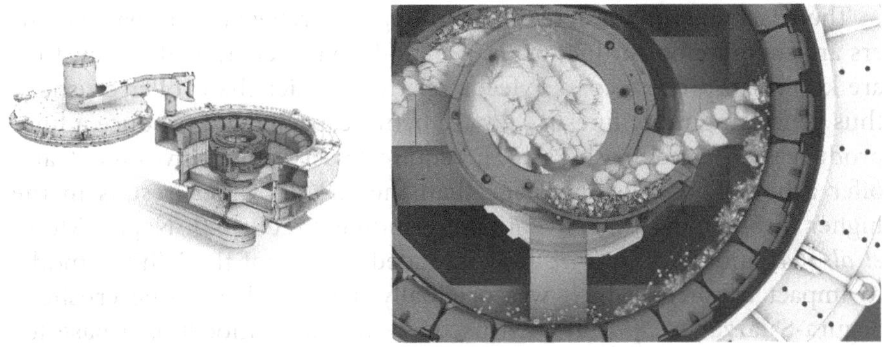

Figure 1.4 VSI [55].

The authors noted that the design of the VSI components significantly affects the productivity of the crusher and thus improves its efficiency through optimizing the impact angle between particles and the impact segment of ring armor. The authors defined the ring armor as 'a circular cylinder around the centrifugal rotor where the high-speed particles that leave the rotor chambers are crushed as they hit the ring armor surface'. The potential for saving energy is high in VSI as observed by Xia *et al.* [57], especially when the equipment is operated efficiently. The efficient operation of the crusher is hinged upon the optimization of the control variables, such as the belt conveyor feed rate, the speed of the crusher rotor, and the power demand-side management techniques such as the time-of-use (TOU) electricity tariff. Among the different rotor circumferential speeds investigated, the crusher was found to operate efficiently between 70 and 80 m/s.

The TOU electricity tariff has drawn the interest of researchers such as Numbi *et al.* [29] who took the energy cost as the performance index to be minimized. They achieved this by controlling the TOU through the variables in the developed model, which are the feed rate of the belt conveyor, feed rate of the VSI crusher rotor, and the cascade flow rate. Interesting to note is that optimizing energy expenditure in the mineral processing industry based on TOU tariff has received attention from different researchers earlier, with Zhang and Xia [58, 59] optimizing the belt conveyor system, while Badenhorst *et al.* [60] used the method to schedule the host operations of a deep level mine twin rock winder system. Numbi *et al.* [29] then later extended it to optimize energy expenditure of the comminution processes. The authors managed to reduce the energy cost by 21.8% through shifting the load from the peak to off-peak and standard periods.

There is a unanimous agreement among researchers that impact crushers are energy-efficient [54, 61–63]. In addition to energy efficiency, they are known to process coarser feeds relatively easier than roller crushers, thus generating more fines [61]; have better capacity and produce a finer product than cone crushers [63]; require lesser capital investment and offer a better pre-crushing stage than the HPGR, which results in the higher throughput of a conventional grinding circuit [62]. Napier-Munn *et al.* [64] and Kojovic *et al.* [65] extended the use of the Whiten model to impact crushers, which was originally developed for cone crushers. Segura-Salazzar *et al.* [54] then used the modified model as a base for modeling the VSI coupled with industrial experiments. They extended the modified model by equations with the K3 parameter in the classification function and T10 parameter in the breakage function, which is the percentage of product passing 1/10th of the original particle size after breakage. A detailed procedure for determining the K3 and T10 parameters is given [54], and the parameters were found to vary with feed rate and rotor speed. The experiments conducted in that work covered a range of feed rates, rotor speeds, feed distributor types, and feed size distributions. From that approach, specific energy consumption, and the product size distribution were confidently predicted over a wide range of operating conditions. An inverse relationship was observed between the feed rate and the proportion of material entering the rotor, whereas a direct relationship was observed between the feed rate and the cascading flow.

Modeling also had its fair share of improving the performance of impact crushers, which benefits current operators. The dawn of DEM has enabled researchers to model the process mechanistically [51–66]. Although the mechanistic models are recent and make meaningful progress, the simple nature, easy implementation, and integration ability with other unit operations of the phenomenological models make optimization of the entire size reduction unit simple. It is important to mention for the benefit of the readers that mechanistic models use theory to predict what happens during the crushing process and the results thereof, whereas phenomenological models describe a hypothesized relationship between variables in the data set, which seeks to best describe the data, yet not derived from the first principle. In light of modeling, Whiten and White [41] described particle breakage using the classification function and particle breakage of selected particles. This thus formed a base on which other researchers built upon to compare mechanisms of volume and surface breakage [67] and propose breakage and classification functions applicable to impact crushers. From such models, the rotor radius and rotation frequency were determined and validated for the HSI crusher [54].

Sinnot and Cleary [51] used DEM to assess the breakage and particle flows in the hammer mill, HSI, and VSI. The authors provided a platform to compare three impact crushers using an energy absorption-based threshold breakage. The method of breakage in these crushers relies on either fracturing slowly moving feed by an impeller moving at high speed or propelling particles at high speed into stationary obstacles. Either way, the feed particles are fractured into progeny based on geometric parameters and estimates of the stress applied to particles flowing through the crusher. From the simulations performed by Sinnot and Cleary [51], key operational parameters essential for crusher performance such as the throughput, power draw, and machine wear, were predicted. DEM was also used to assess breakage for single particle impacts in both VSI and HSI crushers by Djordjevic *et al.* [68].

Still on modeling, Chimwani and Bwalya [18] investigated the influence of particle size and impeller speed on the rate of grinding and product size distribution using experimental work and PBM. Their results demonstrate the ability of the PBM to describe the breakage process and the generation of particle size distribution in the vertical impact crushers. The work also revealed the important role played by fissures in the particles, and that multiple passes are not required for higher rotor speeds due to a decrease in the net grind time, which makes the process inefficient. This is proof that tools such as the DEM and PBM could be useful to optimize the processing of low-grade ores by impact crushers, especially considering the quantitative relationship between the ore crushing characteristics, mineral content, and grindability as observed by Genc and Benzer [69].

1.2.1.4 Cone Crusher

The cone crusher shown in Figure 1.5 is classified as a secondary crusher, lighter than the heavy-duty-rugged primary machine. It is widely used for secondary and tertiary crushing in the mineral processing circuits [70]. The purpose of the crusher is to break particles to sizes suitable for grinding, thus the crusher serves as a tertiary crusher, especially when positioned between the ball mill and secondary crusher in a close circuit [24]. It is important to note, though, that secondary and tertiary crushers are differentiated by the set, which is closer in the latter than the former. This type of crusher is like a gyratory crusher for the eccentric motion of the inner crushing cone, but is differentiated by the spindle, which is shorter and not suspended but rather supported in a curved universal beating below the head or cone of the gyratory head. Additionally, the crushing surface of the cone crusher has a relatively larger diameter than its vertical dimension.

When particles drop into its crushing chamber, the mantle of the crusher gyrates in an eccentric motion, swinging slightly as it rotates. This alters the gap between the concave and mantle, crushing the material at points where the gap is smallest, known as the closed-side setting. Opposite to the smallest gap on the other side is the maximum constricted separation of the concave and mantle, known as the open-side setting (OSS) [71]. The OSS takes control of the vertical flow of material, which affects particle mobility, moving downwards. Thus, the OSS determines the largest particle of the outgoing product, whereas the CSS determines the final product size, the crusher capacity, and the energy consumption of the crusher. The shapes of the gyrating cone and concave, which tend to be altered by wear over time, determine the performance of the cone crushers, hence the need to adjust the mantle upwards every now and then, to maintain the CSS.

The cone crusher has received considerable attention from researchers, which has greatly improved its performance [72–74]. In view of improving the crusher performance and efficiency, Lee and Evertsson [75] investigated the effects of the closed side setting, eccentric speed, and stroke on crusher output, thus comparing the existing cone crusher performance against theoretically determined optimal. Results from this investigation show that the product yield improves as the eccentric speed decreases. This is owing to the relationship between the eccentric speed and the crushing zones or applied compressions inside the crushing chamber, which is directly proportional. The authors also concluded that the product yield is also improved by a larger CSS. Thus, when enlarged, the CSS increases the size of the smallest permitted size as well as a larger stroke, which effect an increase in the amount of the circulating load and a decrease in the amount of fines produced. Cleary *et al.* [71] also regard CSS as the determining factor of the peak strain needed for a given particle to break and the progeny formed. Lee and Evertsson [75] reiterate that smaller compressions are better than what is used in modern crushers, an assertion that supports their earlier work [76]. Such assertions that challenge the status quo of modern crusher operations are welcome, especially during the processing of low-grade ores, which are costly.

To provide more insights into the operation of the cone crusher, Johansson *et al.* [25] used DEM to simulate the operation of the laboratory cone crusher by modeling particle breakage using the bonded particle model approach and validating their simulations with experiments. For two high eccentric speeds investigated, a higher power draw was recorded. The authors also further unpacked the approaches used for modeling particle breakage in DEM and their applicability to modeling compressive breakage in cone crushers. They showed that the bonded particle model approach with spheres arranged in a cluster and bonded together by bonding beams in each contact point was used in the cone crusher simulations by Quist and Evertsson [77, 78]. The particle replacement model in which particles are replaced by a set of progeny fragments produced at the breakage event was utilized by Cleary and Sinnott [79] and Delaney *et al.* [80] to simulate the same crusher. Then, the tetrahedral element model approach that has particles modeled using a tessellated mesh structure with Voronoi grains, polyhedrons, or trigons was used by Lichter *et al.* [81] but on the laboratory cone crusher.

Cleary *et al.* [71] then extended the inquiry of the cone crusher performance using DEM to an industrial scale, varying material properties and operational conditions to assess the breakage behavior and the operating performance of the crusher. Imperative to note in that work is that the combination of concave and mantle (gyrating cone) creates regions in the compression region differentiated by the width and degrees of convergence between the surface. This in turn determines the particle breakage, as well as the source and strength of the forces, fundamental to particle breakage. The models used in the DEM simulations include t_{10} rock breakage-based characterization with non-round particles, which uses a replacement strategy and impact energy specific progeny size data. From that work emerged that the nature of the feed size has no dramatic influence on the behavior of the crusher.

Additionally, Cleary *et al.* [71] also found out that cone crusher performance is degraded by the non-uniformity flow of the product down the lower part of the mantle when larger feed particles jam prior to breakage in the crushing zone, and thus, restricts the flow of material. The authors also found from their simulations that there is no significant improvement in throughput realized in processing mixed hard and soft ore since soft particles will be held up during the deformation of the hard particles. Crushing material of higher strength, which also has implications for the power draw, results in the production of a finer progeny with a generally coarser product PSD. This is due to the severe breakage of particles emanating from being highly loaded before they break. For this crusher, higher throughputs are associated with low particle-particle and particle-liner frictions, which consequently lead to the production of a coarser product, and higher throughput is also an offspring of higher CSS and a larger gap between the mantle and cone. In cases where a higher size reduction ratio is important and maximizing the production of a given size is not, choke feeding is preferred, which promotes rock-on-rock breakage, as observed by Evertsson [82]. Higher rotation speeds shorten the expansion phase of the crusher gap and lengthen the transit time, which eventually has implications of an increased residence time in the cone crusher, hence a finer product at the expense of throughput [71].

Zhang *et al.* [83] established a model based on the population balance model and inter-particle breakage theory, to assess the effect of the relative particle size, particle size distribution, and compression ratios on the inter-particle breakage. The model took into consideration the particle shape transformation, and that enabled the flakiness index of the product from the cone crusher to be determined. Using the model, Zhang and colleagues were able to describe the dynamic process of inter-particle breakage and establish a mathematical relationship between the feed and product as a function of the crushing chamber geometry, granular property of feed, flakiness index of the crushed product, and the cone crusher operational parameters. Gröndahl *et al.* [84] suggested the feed variance across the crushing chamber as the source of power draw fluctuations in a cone crusher. Feed variations are in turn caused by variable feed size distribution, mass flow rate, density variations of the particle beds in the crusher cavity originating from particle segregation, and feeding alignment. Having identified feed segregation, stiffness, and shape of the product as factors that affect crusher performance in their earlier work [84, 85], the authors found the change in attributes of the feed across the geometric position on a belt feeder or in the crusher (segregation) to be very influential to the power draw. Feed segregation is caused by feed with inherently different material, causing smaller particles to fall to the bottom while larger particles rise to the top. The authors also noted that crushing of material in a circular motion causes the feed across the crushing

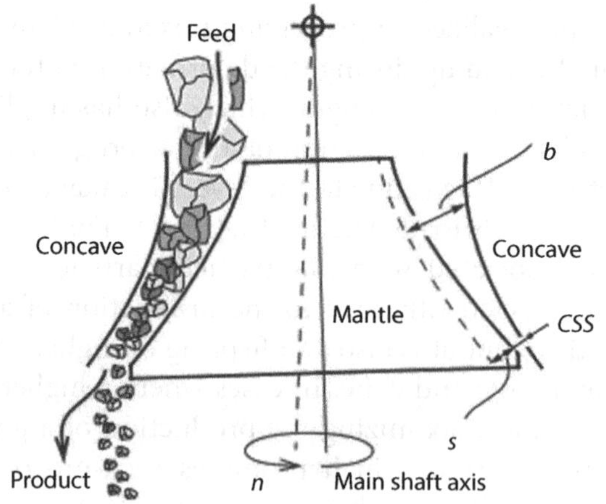

Figure 1.5 Schematic diagram of a cone crusher [82].

chamber to vary and consequently causes fluctuations in power draw. This is mostly influenced by the mass flow and particle size distribution among other material properties.

To provide more information about cone crusher operations, Cleary and Morrison [86] gave a comprehensive assessment of the geometric measures to determine how cone crusher flows and breakage vary with key crusher geometric parameters. The variables assessed are the open-and-closed side settings, geometry-based size reduction ratio, cross-sectional area available for flow at each height, and the eccentric throw, and the assessment was for the specific short head cone. The CSS being the control area for the degree of breakage, determines the product size of any crusher geometry, while the throughput is controlled primarily in the choke area [86]. The authors also found that the variations of the cross-sectional and flow area with height can be used to quantify the geometric constrictions, which control the throughput and flow, with the variations of flow area with height occurring on the CSS side. Other important observations made are as follows: raising the cone reduces the flow and throughput, both the CSS and OSS vary with cone height linearly, and the current state of the configuration of the crusher requires the knowledge of either the OSS or CSS.

In their review of the cone crusher technology and the development and application of process models in crushing circuits optimization, Yamashita *et al.* [87] unpacked the modeling and control strategies for cone crushers in the mineral processing and quarrying industries undertaken in different published works. The review discussed model formulation, which comprises population balance models [88–91], data-driven (empirical) models [92–94] as well as control strategies that comprised model derivative control, proportional-integral-derivative and expert system. Yamashita *et al.* [87] suggested that it is imperative for control strategies to utilize online information from gadgets, such as hyperspectral cameras to improve production rates and optimize crushing energy expenditure. Also, it is important to strike a balance between the crushing performance and liner wear and to use machine learning techniques to improve overall production rates and achieve crushing power consumption optimization. The review also suggested that sensors could be developed in the future so that the pressure and shear stress would be mapped precisely along the whole surface liners, hence developing better power draw and wear models based on mechanistic principles. The behavior of the cone crusher should be completely understood when running empty or under choke-fed conditions and the machine should be able to switch to safe modes if non-crushable foreign objects are detected. Some researchers have identified cone crusher models, which are of a dynamic nature with the ability to assess power consumption, and flow of particles albeit with affordable computational power requirements [89–95]. This would help to adapt cone crushers to low-grade ore processing through integrated planning of mining and mineral processing tasks since the relevance of sustainable processes cannot be overemphasized. Yamashita *et al.* [87] also identified that currently, DEM can only avail the fundamental knowledge of the flow and breakage of rocks inside the cone crusher chamber, but the technique has the potential to become a powerhouse for modeling cone crushers, especially with the rise of computing power and more efficient algorithms.

Still, on modeling of cone crushers, Lindqvist and Li [96] proposed a model formulated within the PBM framework to assess the influence of CSS on the product size distribution. This enabled the amount of energy required to produce a given product size distribution to be calculated. The authors suggested that since there are correlations among the CSS, feed top size, chamber length, mantle size, and eccentric throw, inputting some of these parameters and not all in the model would be adequate to achieve the intended purpose. Some of the findings observed a direct proportionality relationship between the Bond work index of the rock and the CSS. Thus, the Work index influences the CSS, and the corresponding particle size distribution produced. Appreciating the potential of the feeding mode to influence the cone crusher performance, Nematollahi *et al.* [97] investigated the effect of a feed chute on the crusher using DEM. This study was motivated by the assertion of Gröndahl *et al.*, [84] that poor feed distribution is the potential source of high-power draw fluctuations in crushers, hence inefficiencies of the process. With the new design proposed, Nematollahi *et al.* [97] were able to improve the crusher throughput by 36%, and achieved a narrower and finer product size distribution as well as the uniform and decreased wear rate on the liners and mantle of the crusher.

André and Tavares [98] also contributed towards the understanding of the cone crusher operation by applying a breakage model fitted using single-particle breakage information and polyhedral particles. The parameters at the center of their investigation were CSS, stroke, frequency, and throughput. What the authors found in this work generally agrees with literature and industrial practice, that CSS is directly proportional to throughput but inversely proportional to specific energy. Product fineness and throughput also increased with the crusher throw. As for the feed blends, a small rise in the proportions of tough material in the feed linearly increases specific energy, albeit with a significant drop in throughput. These findings led to the conclusion that the combination of a breakage model and a particle replacement scheme with polyhedra is adequate to simulate a cone crusher. The breakage model, thus, has to capture the fracture energy variability of the feed.

1.2.1.5 HPGR Crusher

High-pressure grinding rolls (HPGR) utilized in the secondary and tertiary crushing application of mostly friable, frozen, and less abrasive feeds have either one rotating cylinder or a pair of movable cylindrical manganese steel rollers rotating in opposite directions with adjustable speeds to break the particles. The particles are broken by means of compressive forces, producing a cake with a high percentage of fine materials. Compression forces in the compacted bed lead to the formation of microcracks in feed particles, causing them to break along natural crystal lines and exit the crusher through the discharge point. The discharge point depends on the set, which is determined by shims, which are responsible for holding back the spring-loaded roll from the solidly mounted roll. If the set is smaller, a generally finer product is produced but at a high-operating pressure of the rollers, which is directly proportional to the energy input of the HPGR. Roller speed and compressive pressure have been found to be influential to the shape and size of the product of the crusher [99], which are mostly angular and elongated. Particles, through the crusher, experience compressive pressure only once, contrasting the gyratory and jaw crushers in which the compression pressure is repeated in a progressive size reduction fashion as the material approaches the discharge point. Although they require a high capital cost, the crusher provides a high size reduction ratio with an improved degree of mineral liberation in the crushed particles and its energetic benefits are indisputable [100]. The material loading in the HPGR zone among other factors has made the crusher more energy-efficient, producing a product with higher fines compared to the impact and compressive crushers.

Considerable work has been done to model and optimize the operation of HPGR crushers by different researchers [101]. This helped unlock their potential processing benefits such as the non-selective crushing of coarse particles [102], particle weakening, improved liberation owing to microcrack formation, and energy savings [103, 104]. Having realized the need to quantify particle damage and downstream benefits from HPGR crushing, Kodali *et al.* [104] assessed particle damage and copper grain exposure of the jaw crusher and HPGR crusher products from crushing copper ores using X-ray computed tomography. From their comparison of the two crushers, the authors found that an HPGR produces more damaged particles with high grain exposure, which led to high copper recoveries during leaching.

They attributed this to higher inter-particle stresses generated when the set of pressurized rollers compresses the particles as they move through the narrow gap between the rollers. The HPGR also produces particles with microcracks [105], which promotes more penetration of leach solution to solubilize the mineral grains, hence high metal recovery and lower downstream requirements. However, the inability of the crusher to handle metalliferous ores [106] makes the crusher performance ore texture dependent. For both copper oxide and sulfide samples, Kodali and colleagues found that the HPGR produces finer particle size distribution, more internal cracks in the majority of the particle size classes, products with faster grain dissolution, and higher copper recovery from leaching copper oxide. Nonetheless, copper sulfide leaching results suggest that high recoveries are not sensitive to crusher types in spite of the greater particle damage from HPGR [107].

Inspired by the inroads and progress made by HPGR into the mineral processing industry, Numbi and Xia [36] demonstrated that it is possible to use a systems optimization model to manage the energy of a parallel HPGR crushing process. The system optimization model enabled them to reduce energy consumption by optimally controlling the process and reducing the cost through load shifting. Systems optimization works as a supervisory control system at the upper control layer by minimizing energy cost, emissions, and downtime or maximizing energy efficiency, total plant throughput, and product quality. Assuming equal overall drive efficiency and fixed rolls operating pressure, the optimization scheme of Numbi and Xia [36] revealed a potential to reduce specific energy cost without reducing energy consumption. In a scenario where the operating pressure of the rolls was fixed but the drive efficiency of the two parallel HPGR was variable, energy was saved by minimizing the less-efficient HPGR machine load level and vice-versa for the more efficient HPGR machine. Their results also demonstrated that the product quality is less affected than the specific energy consumption of the HPGR by the change in roll operating pressure.

Genc and Benzer [107] studied the effect of pre-grinding with HPGR on the grinding capacity of an industrial classification circuit composed of a two-compartment cement ball mill. This was following the suggestion of Ghosh [108] that an HPGR can be used as a pre-grinding unit before a ball mill, operated in a hybrid finish and semi-finish grinding closed circuits. Results from the work of Genc and Benzer [107] showed an increased throughput and a finer particle size distribution inside the mill when the clinker crushed with HPGR was used. These kinds of investigations, in which the performance of a ball mill is related to the crushed product upstream, are necessary to determine the optimal processing conditions of low-grade ores, which have lower profit margins than high-grade ores.

In another work, Ballantyne *et al.* [109] used the Bond mill test procedure to investigate the downstream milling energy requirements for HPGR, specifically focusing on the ball mill requirements. Among other findings, the authors observed that HPGR crushing results in weaker coarser progeny owing to microcracking and intact fines. The authors, however, warned that the effect of the changes in the shape of the size distribution should not be related to the rock strength when characterizing the product of HPGR for ball milling. The microcracks observed by Ballantyne *et al.* [109] were further investigated by Yin *et al.* [110] to determine their influence on the leaching of gold ore for HPGR and jaw-crushed ore. The larger fractal dimension observed for the HPGR crusher suggests that it produced more suitable samples with higher proportions of fines, more regular shape with richer microcracks and less uniformity than the jaw crusher. Particles with more microcracks produced enhanced the infiltration of the leaching reagent and consequently high gold recovery in full-slime and cyanide leaching, consolidating observations of other researchers about the effect of microcracks on coarse particles surfaces [109–113]. Yin *et al.* [110] assert that higher and faster gold extractions can be achieved with the use of HPGR. The chemical, physical, and surface properties of HPGR have not received more attention compared to the energy consumption of the machine.

Cleary and Sinnot [114] utilized the DEM to investigate the axial pressure distribution, flow behavior, breakage, and stress distribution within the HPGR crusher with smooth rolls and cheek plates. This is owing to the held belief that particles closer to the ends of the rolls are less well confined with high mobility, which reduces their ability to be trapped and crushed, thus negatively affecting the performance of the crusher [114]. Results from this work showed two mechanisms for generating particle breakage in this crusher, the balance of which depends on the strength of the rock. The first mechanism is breakage near the roll surfaces generated by high shear stresses in the frictional boundary layers from the high peripheral speed of the rolls immediately adjacent to them. The ability of this mechanism to break rocks is inversely proportional to the rock strength. In the second mechanism, which becomes more significant with the increase of the strength of the rock, the regions where the frictional boundary layers converge and around the pinch point generate high normal stresses. In addition to that, particles are subjected to very high compression that generates breakage in those regions. The results also revealed that the reduced bed pressure decreases breakage near the ends of the rolls, leading to a slight increase in the product size with an increased displacement of the cheek plates.

In their flake graphite upgrading process based on HPGR, stirred grinding mill, and nanobubble flotation, Ma *et al.* [115] found that better performance of the flotation process was attributed to an efficient and selective comminution process achieved using an HPGR. The authors also observed that HPGR generates cracks along the interface between graphite and gangue minerals, which further promotes the liberation and protection of flakes in the successive grinding process. In the same year, Yin and Chen [116] assessed the characteristics of particle size and microstructure of the product from HPGR on the bioleaching of agglomerated copper ore. This was necessitated by the negative effect of the excessively fine particles produced by HPGR on heap permeability, thus limiting its application in heap leaching. Their results show that the HPGR produced particles with more microcracks and high porosity and heap permeability in the column was affected by the excessive fines, confirming earlier observations [117, 118]. But the leaching efficiency in the early stage of column leaching was higher for material from an HPGR crusher, confirming earlier observations that comminuting the ore with HPGR effectively improves heap leaching efficiency [104, 110]. It was also shown that agglomerated particles from the HPGR crusher could be efficiently leached owing to its particle size and microstructure characteristics, thus providing a new way of efficiently leaching low-grade copper ore [116].

The population balance model was also not left out in the analysis of particle breakage in the HPGR crusher [119]. The authors assessed the generation of progeny particles by the comminution process of tantalum and tungsten ores using a piston-die press, and it gave insights into the material response in an HPGR. Their results showed a normalized breakage distribution for a single particle compression and a non-normalized breakage distribution for a bed of particle compression, which implies that a single particle and a bed of particles have different breakage functions. Using a methodology established in their earlier work, Anticoi *et al.* [120] used an improved approach to analyze different breakage distribution function parameters for a single particle and a bed of particles. They, however, used the same specific pressing force with variable feed particle size, static gap, and roll speed, and found that varying rotational speed at a constant feed size range increased the throughput considerably without influencing the particle size distribution. Using a piston-and-die test, Campos *et al.* [121] investigated the breakage of confined beds for narrow sizes of fine iron ore concentrates, and the results revealed that the breakage of top size fraction reached a maximum at high pressures and the daughter fragments from the parent particles generates additional fines. Results also showed that part of the inelastic energy does not contribute to the generation of new surfaces but is rather utilized in dissipative processes, such as plastic deformation, friction, and particle packing.

Table 1.1 Comparison of different types of crushers.

Type of crusher	Feed (m)	Type of force	Stage	Product size determining parameter
Jaw crusher	1.5	Compressive	Primary	CSS
Gyratory crusher	1.37	Compressive	Primary	Set (gap setting)
Impact crusher	0.6	Impact	Secondary/ Tertiary	Rotor liner clearance
Cone crusher	0.432	Compressive	Secondary/ Tertiary	CSS
HPGR	0.15	Compressive	Secondary/ Tertiary	Roller gap

Barrios *et al.* [35] investigated the effect of roller aspect ratio, lateral confinement, and wear condition on the performance of the HPGR crusher. The authors used the combination of DEM, multi-body dynamics, and the particle replacement model to simulate a pilot-scale HPGR grinding pellet feed. Their results showed that the studs fitted rotating side plates resulted in more uniform forces, which promote the production of constant product size along the rolls at higher throughput. The edge effect was also shown to be independent of the roll length, resulting in higher proportional edge regions for high-aspect-ratio rolls. The worn rolls were observed as being problematic for HPGR [122], owing to their generation of heterogeneous force profiles along the rolls with the bathtub-like profile of both forces producing a coarser product with a trapezoidal profile. As stated by Barrios *et al.* [35], increasing the roll speed and specific force, operating at the above minimum gap, often compensates for roller wear to maintain the throughput and product fineness.

Table 1.1 compares different crushers in terms of the feed particle size they can handle, type of the force used to crush the particles, the crushing stage at which they are mostly utilised and the product size determining parameter.

1.2.2 Gaps and Future Perspective

Different crushers (jaw, gyratory, cone, impact, and HPGR) were reviewed in this chapter to highlight distinct efforts put in by different researchers.

This helps to understand their mode of operation and achieve optimum operation of those size reduction machines. The optimum operation is achieved by controlling the parameters that influence their operation so that they can operate at high capacity and low cost. Increasing the capacity and reducing the operational costs of the crushers consequently improve the performance of the comminution circuit. This is because the efficient operation of the crushing circuit means that the 'more crushing and less grinding' concept can be implemented in comminution circuits to improve their processing capabilities and reduce the energy consumption of the plant. This is very crucial, especially considering the domination of low-grade ore processing brought about by the depletion of high-grade ores.

For the jaw crusher, the swing jaw pivoting position, which differentiates jaw crushers, determines the type of ores each type of jaw crusher efficiently crushes. In that vein, single toggle crushers were found suitable for soft ores, while double toggle is suitable for tough and abrasive ores. It is common knowledge that despite variations in ore feeds over the years, crushers installed at the inception of the mining activities are still operational. This does not accommodate the deterioration in the ore grades of the ore processed over time, let alone the mechanical strength of feed ores caused by different uniaxial compressive strengths in different rocks, which negatively affects the efficiency of the crushing circuits. The behavior of rock aggregates under choke feeding tests also revealed that the efficiency of each crusher is hinged on its breakage mechanism. Thus, to reduce processing costs, ores should be processed by suitable crushers to maintain efficiency. It was also observed that compressive crushers, such as jaw crushers, not only lose a significant portion of energy into noise and heat, but have no-load power ranging from 40% to 50%. Therefore, optimal switching control techniques were found handy in reducing the energy consumption of these crushing machines. This chapter also revealed that none of the studies undertaken about jaw crushers to date were downstream performance-oriented, but the studies were rather focused on improving the efficiency of the crusher itself.

In gyratory crushers, the crushing chamber was found to have a potent influence on the quality and cost of the subsequent downstream processes. A case in point was the purification of iron and such downstream-oriented studies, though insightful and needed in the wake of low-grade ore processing, are still sparse in the literature. It was revealed that optimizing the operation of a gyratory crusher is impossible without taking into consideration the properties of the feed, such as the work indices and the fines content. Thus, further research in that direction would improve the performance of crushers with the consequence of lowering the extraction cost of minerals. Important avenues, which have not been explored widely yet have great potential for reducing energy expenditure in gyratory crushers, are optimizing parameters, such as the idling time since the energy expenditure of the machine during idling is 30% of the full load power consumption.

Considerable work has been done on impact crushers experimentally and using different modeling techniques. But what has not received wide attention is the effect of the amount of feed, rotor speed, the mechanism of size reduction in the crusher (determined by the velocity and energy distribution of collisions inside the crushing chamber), the lining design, and the design of the hammer on the impact crusher. The power demand-side management techniques, such as the TOU electricity tariff, have proved to be effective in reducing the energy consumption of impact crushers by shifting the load from the peak to off-peak and standard periods. This should be common practice, especially during low-grade ore processing, to reduce the processing costs significantly. With the increase in computational power of modern machines, modeling techniques such as the DEM are expected to provide the much-needed guidance for efficient low-grade ore processing, by impact and other types of crushers.

In the future, smaller compressions than what are used in modern cone crushers are expected to improve the performance of such crushers. Researchers should therefore carry out more studies in that space to validate such claims. Claims that challenge the status quo of modern crusher operations are welcome, especially during the processing of low-grade ores, which are costly. DEM was tipped to have the potential to become a powerhouse for modeling cone crushers, especially with the rise of computing power and more efficient algorithms.

For HPGR, research work has shown that the chemical and surface properties of these crushers have not received more attention compared to the energy consumption of the machine. As observed for the crushers and according to the discussions in the foregoing, investigations that relate downstream units, such as ball mills to the crushers upstream, are necessary to determine the optimal comminution circuit processing conditions. This is even much needed for low-grade ore processing, which results in lower profit margins than high-grade ores.

1.3 Conclusion

This chapter reviews different crushers to unlock their potential to process different ores efficiently. This is in pursuance of the more crushing and less grinding concept brought about by the realization that crushers are 10 to 20 times more energy-efficient compared to milling. From the review of crushers in this chapter emerged that more downstream processes-oriented research, rather than those that focus on the optimization of a particular unit, are needed to improve the performance and reduce the operational cost of crushers. Different avenues with great potential to reduce the energy consumption of different crushers were considered. This includes managing the demand-side of power, such as the TOU electricity tariff technique that has proved to be effective in reducing the energy consumption of impact crushers. Optimizing the idling time of the crushers also helps to reduce energy consumption since the energy expenditure of the gyratory crusher during idling is 30% of the full load power consumption, and jaw crushers not only lose a significant portion of energy into noise and heat, but have no-load power ranging from 40% to 50%.

References

1. Seccatore, J., A review of the benefits for comminution circuits offered by rock blasting. *REM*, 72, 1, 2019. http://dx.doi.org/10.1590/0370-44672017720125.
2. Napier-Munn, T., Is progress in energy-efficient comminution doomed? *Miner. Eng.*, 73, 1–6, 2014.
3. Legendre, D. and Zevenhoven, R., Assessing the energy efficiency of a jaw crusher. *Energy*, 74, 2014.
4. Masamuda, H., Higashitani, K., Yoshida, H., *Powder Technology Handbook*, 3rd ed, Boca Raton (FL, Taylor and Francis Group), 2006.
5. Richardson, J.F. and Harker, J.H., Particle technology and separation processes, in: *Coulson And Richardson's Chemical Engineering*, 5th ed, vol. 2, England, Elsevier Science, Oxford, 2002.
6. Gupta, A. and Yan, D.S., Chapter 4-Jaw crusher, in: *Mineral Processing Design and Operation*, A. Gupta and D.S. Yan (Eds.), pp. 99–127, Elsevier Science, Amsterdam, 2006.
7. Tromans, D., Mineral comminution: Energy efficiency considerations. *Miner. Eng.*, 21, 613–620, 2008.
8. Cundall, P.A. and Strack, O.L., A distinct element model for granular assemblies. *Geotechnique*, 29, 47–65, 1979.
9. Mishra, B.K. and Rajamani, R.K., Motion analysis in tumbling mills by the discrete element method. *KONA Powder Part. J.*, 8, 92–98, 1990.
10. Cleary, P.W., Modelling comminution devices using DEM. *Int. J. Numer. Anal. Methods Géoméch.*, 25, 83–105, 2001.

11. Weerasekara, N., Powell, M., Cleary, P.W., Tavares, L.M., Evertsson, M., Morrison, R., Quist, J., Carvalho, R., The contribution of DEM to the science of comminution. *Powder Technol.*, 248, 3–24, 2013.
12. Tavares, L.M., A Review of Advanced Ball Mill Modelling. *KONA Powder Part. J.*, 34, 106–124, 2017.
13. Cunha, E.R., Carvalho, R.M., Tavares, L.M., Simulation of solids flow and energy transfer in a vertical shaft impact crusher using DEM. *Miner. Eng.*, 43-44, 85–90, 2013.
14. Chimwani, N. and Bwalya, M.M., Exploring the end-liner forces using DEM software. *Minerals*, 10, 12, 1047, 2020.
15. Bwalya, M.M. and Chimwani, N., Numerical simulation of a single and double-rotor impact crusher using discrete elemental methods. *Minerals*, 12, 2, 143, 2022.
16. Shi, F., Kojovic, T., Esterle, J., David, D., An energy-based model for swing hammer mills. *Int. J. Miner. Process.*, 71, 147–166, 2003.
17. Chimwani, N., Mulenga, F.K., Glasser, D., Hildebrandt, D., Bwalya, M., Use of the attainable region method to simulate a full-scale ball mill with a realistic transport model. *Miner. Eng.*, 73, 116–123, 2014.
18. Chimwani, N. and Bwalya, M.M., Milling studies in an impact crusher i: Kinetics modelling based on population balance modelling. *Minerals*, 11, 470, 2021. https://doi.org/10.3390/min11050470.
19. Chimwani, N., A review of the milestones reached by the attainable region optimisation technique in particle size reduction. *Minerals*, 11, 1280, 2021. https://doi.org/10.3390/min11111280.
20. Legendre, D., *Assessing the Energy Efficiency Of Jaw Crushers*, Åbo Akademi University [MSc (Eng) Thesis], Turku, Finland, 2012.
21. Kaya, E., Hogg, R., Kumar, S., Particle shape modification in comminution. *KONA Powder Part. J.*, 20, 188–195, 2002.
22. Holt, C.B., The shape of particles produced by comminution-a review. *Powder Technol.*, 28, 59–63, 1981.
23. Durney, T.E. and Meloy, T.P., Particle shape effects due to crushing method and size. *Int. J. Miner. Process.*, 16, 109–123, January 1986.
24. Haldar, S.K., Chapter 13-Mineral processing, in: *Mineral Exploration*, Second Edition, S.K. Haldar (Ed.), pp. 259–290, Elsevier, 2018.
25. Johansson, M., Bengtsson, M., Evertsson, M., Hulthén, E., A fundamental model of an industrial-scale jaw crusher. *Miner. Eng.*, 105, 69–78, 2017.
26. Gauldie, K., *The Performance of Jaw Crushers, Engineering*, October 9, 1953.
27. Oduori, M.F., Mutuli, S.M., Munyasi, D.M., Analysis of the single toggle jaw crusher kinematics. *J. Eng. Des. Technol.*, 13, 2, 213–239, 2015.
28. Tosun, A. and Konak, G., Development of a model estimating energy consumption values of primary and secondary crushers. *Arab. J. Geosci.*, 8, 2, 1133–1144, 2015.
29. Numbi, B.P., Xia, X., Zhang, J., Optimal energy control modelling of a vertical shaft impact crushing process. *Energy Procedia*, 61, 560–563, 2014.

30. Fladvad, M. and Onnela, T., Influence of jaw crusher parameters on the quality of primary crushed aggregates. *Miner. Eng.*, 151, 106338, 2020.
31. Olaleye, B.M., Influence of some rock strength properties on jaw crusher performance in granite quarry. *Min. Sci. Technol.*, 20, 0204–0208, 2010.
32. Köken, E. and Lawal, A.I., Investigating the effects of feeding properties on rock breakage by jaw crusher using response surface method and gene expression programming. *Adv. Powder Technol.*, 32, 1521–1531, 2021.
33. Nikolov, S., A performance model for impact crushers. *Miner. Eng.*, 15, 715–721, 2002.
34. Donovan, *Fracture Toughness-Based Models For The Prediction Of Power Consumption, Product Size, And Capacity Of Jaw Crushers*, Mining and Minerals Engineering Department, Virginia Polytechnic Inst, Blacksburg, Dissertation, 2003.
35. Barrios, G.K., Jimenez-Herrera, N., Natalia, F.-T., Tavares, L.M., DEM simulation of laboratory-scale jaw crushing of a gold-bearing ore using a particle replacement model. *Minerals*, 10, 717, 2020.
36. Numbi, B.P. and Xia, X., Systems optimization model for energy management of a parallel HPGR crushing process. *Appl. Energy*, 149, 133–147, 2015.
37. saVRee. https://savree.com/en/encyclopedia/gyratory-crusher. Accessed on 01 August 2022.
38. Franklin, J., A. and Dusseault, M.B., *Rock Engineering*, McGraw Hill Publishing Company, New York, 2009.
39. Korman, T., Bedekovic, G., Kujundzic, T., Kuhinek, D., Impact of physical and mechanical properties of rocks on energy consumption of jaw crusher. *Physicochem. Probl. Miner. Process.*, 51, 2, 461–475, 2015.
40. Comakli, R. and Cayirli, S., A correlative study on textural properties and crushability of rocks. *Bull. Eng. Geol. Environ.*, 78, 3541–3557, 2019.
41. Gupta, A. and Yan, D.S., Chapter 5-Gyratory and cone crusher, in: *Mineral Processing Design and Operations*, Second Edition, A. Gupta and D. Yan (Eds.), pp. 153–168, Elsevier, Amsterdam, 2016.
42. Rosario, P.P., Hall, R.A., Maijer, D.M., Liner wear and performance investigation of primary gyratory crushers. *Miner. Eng.*, 17, 1241–1254, 2004.
43. Gang, D., Fan, X., Huang, D., Analysis and optimization of cone crusher performance. *Miner. Eng.*, 22, 1091–1093, 2009.
44. Chen, Z., Wang, G., Xue, D., Bi, Q., Simulation and optimization of gyratory crusher performance based on the discrete element method. *Powder Technol.*, 376, 93–103, 2020.
45. Li, Q., Gong, Y.D., Song, W.G., Gyratory crusher productivity analysis based on kinematic characteristics of materials. *China Mech. Eng.*, 27, 2034–2039, 2016.
46. Chen, Z., Wang, G., Xue, D., Cui, D., Simulation and optimization of crushing chamber of gyratory crusher based on the DEM and GA. *Powder Technol.*, 384, 36–50, 2021.

47. Andersen, J.S. and Napier-Mann, T.J., *Min. Eng.*, 3, 1/2, 105, 1990.
48. Whiten, W.J., The simulation of crushing plants with models developed using multiple spline regression. *J. South Afr. Inst. Min. Metall.*, 257–264, 1972.
49. Wills, B.A. and Finch, J.A., Chapter 6-Crushers, in: *Wills' Mineral Processing Technology*, Eighth Edition, B.A. Wills and J.A. Finch (Eds.), pp. 123–146, Butterworth-Heinemann, Boston, 2016a.
50. Deniz, V., A new size distribution model by t-family curves for comminution of limestones in an impact crusher. *Adv. Powder Technol.*, 22, 761–765, 2011.
51. Sinnott, M.D. and Cleary, P.W., Simulation of particle flows and breakage in crushers using DEM: Part 2–impact crushers. *Miner. Eng.*, 74, 163–177, 2015.
52. Metso, Nordberg NP Series impact rushers, 2019. [Online] Available at: https://www.metso.com/products/crushers/impact-crushers/horizontal-shaft-impact-hsi-crushers/ [Accessed 10 July 2019].
53. Bracey, R., Weerasekara, N., Powell, M., Performance evaluation of the novel multi-shaft mill using DEM modelling. *Miner. Eng.*, 98, 251–260, 2016.
54. Segura-Salazar, J., Varios, G., Rodrigues, V., Tavares, M., Mathematical modeling of a vertical shaft impact crusher using the Whiten model. *Miner. Eng.*, 111, 222–228, 2017.
55. BHS-Sonthofen, *RSMX-VSI Rotor Centrifugal Crusher*, BHS-Sonthofen, Sonthfen, 2019, https://dam.bhs-sonthofen.de/pinaccess/showpin.do?pinCode=pZaauLQnStJL. https://dam.bhs-sonthofen.de/pinaccess/showpin.do?pinCode=0xBBgLbhTc2O.
56. Al-Khasawneh, Y., Development and testing of a novel mathematical-physical model for the design of ring armor for the vertical shaft impact crushers. *Miner. Eng.*, 170, 106994, 2021.
57. Xia, X., Zhang, J., Cass, W.J., Energy management of commercial buildings-a case study from a POET perspective of energy efficiency. *J. Energy South. Afr.*, 23, 1, 23–31, 2012.
58. Zhang, S. and Xia, X., Optimal control of operation efficiency of belt conveyor systems. *Appl. Energy*, 87, 1929–1937, 2010.
59. Zhang, S. and Xia, X., Modeling and energy efficiency optimization of belt conveyors. *Appl. Energy*, 88, 3061–3071, 2011.
60. Badenhorst, W., Zhang, J., Xia, X., Optimal hoist scheduling of a deep level mine twin rock winder system for demand side management. *Electr. Power Syst. Res.*, 81, 1088–1095, 2011.
61. Sandvik, K.L., Fagerli, S., Frost, T., Change from pressure crushing in roller mills to impact crushing in production of mineral sand. *Powder Technol.*, 105, 436–442, 1999.
62. Jankovic, A., Valery, W., Davis, E., Cement grinding optimisation. *Miner. Eng.*, 17, 1075–1081, 2004.
63. Lindqvist, M., Energy considerations in compressive and impact crushing of rock. *Min. Eng.*, 21, 631–641, 2008.

64. Napier-Munn, T.J., Morrell, S., Morrison, R.D., Kojovic, T., *Mineral Comminution Circuits: Their Operation and Optimization. JKMRC Monograph Series in Mining and Mineral Processing*, 1996.
65. Kojovic, T., Napier-Munn, T.J., Andersen, J.S., Modeling cone and impact crushers using laboratory determined energy-breakage functions, in: *Comminution Practices*, pp. 187–194, 1998.
66. Cunha, E.R., Carvalho, R.M., Tavares, L.M., A predictive model of the vertical shaft impact (VSI) crusher, in: *Comminution '14*, Cape Town, pp. 1–12, 2014.
67. Bengtsson, M. and Evertsson, M., Modeling of output and power consumption in vertical shaft impact crushers. *Int. J. Miner. Process.*, 88, 18–23, 2008.
68. Djordjevic, N., Shi, F.N., Morrison, R.D., Applying discrete element modelling to vertical and horizontal shaft impact crushers. *Miner. Eng.*, 16, 983–991, 2003.
69. Genc, O. and Benzer, A.H., Single particle impact breakage characteristics of clinkers related to mineral composition and grindability. *Miner. Eng.*, 22, 1160–1165, 2009.
70. Lindqvist, M. and Evertsson, C., Improved flow- and pressure model for cone crushers. *Miner. Eng.*, 17, 1217–1225, 2004. https://doi.org/10.1016/j.mineng.2004.06.028.
71. Cleary, P., Sinnott, M., Morrison, R., Cummins, S., Delaney, G., Analysis of cone crusher performance with changes in material properties and operating conditions using DEM. *Miner. Eng.*, 100, 49–70, 2017. https://doi.org/10.1016/j. mineng.2016.10.005.
72. Moshgbar, M., Bearman, R.A., Parkin, R., Optimum control of cone crushers utilizing an adaptive strategy for wear compensation. *Miner. Eng.*, 8, 4/5, 367–376, 1995.
73. Bearman, R.A. and Briggs, C.A., The active use of crushers to control product requirements. *Miner. Eng.*, 11, 9, 849–859, 1998.
74. Hulthén, E., *Real Time Optimization of Rock Crushers*, Thesis for Licentiate of Engineering, Chalmers University of Technology, 2008.
75. Lee, E. and Evertsson, C.M., A comparative study between cone crushers and theoretically optimal crushing sequences. *Miner. Eng.*, 24, 188–194, 2011.
76. Lee, E. and Evertsson, C.M., Optimized compressive crushing with respect to energy consumption, pressure and multiple product yield, in: *Proceedings of the European Symposium of Comminution and Classification 2009*, Espoo, Finland, 2009.
77. Quist, J.C.E. and Evertsson, C.M., Simulating capacity and breakage in cone crushers using DEM. *Comminution 10'*, Capetown, South Africa, 2010.
78. Quist, J. and Evertsson, C.M., Cone crusher modelling and simulation using DEM. *Miner. Eng.*, 85, 92–105, 2016.

79. Cleary, P.W. and Sinnott, M.D., Simulation of particle flows and breakage in crushers using DEM: Part 1–compression crushers. *Miner. Eng.*, 74, 178–197, 2015.
80. Delaney, G.W., Morrison, R.D., Sinnott, M.D., Cummins, S., Cleary, P.W., DEM modelling of non-spherical particle breakage and flow in an industrial scale cone crusher. *Miner. Eng.*, 74, 112–122, 2015.
81. Lichter, J., Lim, K., Potapov, A., Kaja, D., New developments in cone crusher performance optimization. *Miner. Eng.*, 22, 7, 613–617, 2009.
82. Evertsson, C.M., *Cone Crusher Performance*, Ph.D. Thesis, Dept. of Machine and Vehicle Design. Chalmers University of Technology, Sweden, 2000.
83. Zhang, Z., Ren, T., Cheng, J., Jin, X., The improved model of inter-particle breakage considering the transformation of particle shape for cone crusher. *Miner. Eng.*, 112, 11–18, 2017.
84. Gröndahl, A., Asbjörnsson, G., Hulthén, E., Evertsson, M., Diagnostics of cone crusher feed segregation using power draw measurements. *Miner. Eng.*, 127, 15–21, 2018.
85. Gröndahl, A., Asbjörnsson, G., Quist, J., Hulthén, E., Evertsson, M., Impact of feeding position on power draw size reduction in a cone crusher, in: *European Symposium on Comminution and Classification*, Izmir, 2017.
86. Cleary, P.W. and Morrison, R.D., Geometric analysis of cone crusher liner shape: Geometric measures, methods for their calculation and linkage to crusher behaviour. *Miner. Eng.*, 160, 106701, 2021.
87. Yamashita, A.S., Thivierge, A., Euzebio, T.A.M., A review of modeling and control strategies for cone crushers in the mineral processing and quarrying industries. *Miner. Eng.*, 170, 107036, 2021.
88. Albuquerque, K., Silva, H., Teodoro, E., Fonseca, A., Garioli, G., Lopes, E., Cota, L., Euzebio, T., Averaging level control of bulk solid material using a tripper car. *IFAC-PapersOnLine*, 52, 147–152, 2019.
89. Atta, K.T., Euzebio, T., Ibarra, H., Moreira, V.S., Johansson, A., Extension, validation, and simulation of a cone crusher model. *IFAC-PapersOnLine*, 52, 1–6, 2019. https://doi.org/10.1016/j.ifacol.2019.09.154.
90. Santos, M.S., Pinto, T.V., Júnior, E.L., Cota, L.P., Souza, M.J., Euzebio, T.A., Simheuristic-based decision support system for efficiency improvement of an iron ore crusher circuit. *Eng. Appl. Artif. Intell.*, 94, 103789, 2020.
91. Karlovic, P., Putz, E., Cipriano, A., A framework for hybrid model predictive control in mineral processing. *Control Eng. Pract.*, 40, 1–12, 2015. https://doi.org/10.1016/j.conengprac.2015.02.006.
92. Li, H., Evertsson, C.M., Lindqvist, M., Hulthen, E., Asbjornsson, G., Dynamic modeling and simulation of a SAG mill-pebble crusher circuit by controlling crusher operational parameters. *Miner. Eng.*, 127, 98–104, 2018. https://doi.org/10.1016/j.mineng.2018.07.010.

93. Bhadani, K., Asbjornsson, G., Hulthen, E., Evertsson, C.M., Application of multidisciplinary optimization architectures in mineral processing simulations. *Miner. Eng.*, 128, 27–35, 2018. https://doi.org/10.1016/j.mineng.2018.08.029.
94. Bengtsson, M., Hulthen, E., Evertsson, C.M., Size and shape simulation in a tertiary crushing stage, a multi objective perspective. *Miner. Eng.*, 77, 72–77, 2015. https://doi.org/10.1016/j.mineng.2015.02.015.
95. Quist, J., *Cone Crusher Modelling and Simulation*, Master's Thesis Chalmers University of Technology, 2012.
96. Lindqvist, M. and Li, H., A cone crusher model with a coupled energy- and breakage dependency. *Miner. Eng.*, 172, 107147, 2021.
97. Nematollahi, E., Zare, S., Maleki-Moghaddam, M., Ghasemi, A., Ghorbani, F., Banisi, S., DEM-based design of feed chute to improve performance of cone crushers. *Miner. Eng.*, 168, 106927, 2021.
98. André, F.P. and Tavares, L.M., Simulating a laboratory-scale cone crusher in DEM using polyhedral particles. *Powder Technol.*, 2019. https://doi.org/10.1016/j.powtec.2020.06.016.
99. Abazarpoor, A., Halali, M., Hejazi, R., Saghaeian, M., HPGR effect on the particle size and shape of iron ore pellet feed using response surface methodology. *Miner. Process. Extr. Metall.*, 127, 1, 40–48, 2018.
100. Saramak, D., Challenges in raw material treatment at the mechanical processing stage. *Minerals*, 11, 940, 2021. https://doi.org/10.3390/min11090940.
101. Dundar, H., Benzer, H., Aydogan, N., Application of population balance model to HPGR crushing. *Miner. Eng.*, 50-51, 114–120, 2013.
102. Tavares, L.M., Particle weakening in high-pressure roll grinding. *Miner. Eng.*, 18, 651–657, 2005.
103. Celik, I.B. and Oner, M., The influence of grinding mechanism on the liberation characteristics of clinker minerals. *Cem. Concr. Res.*, 36, 3, 422–427, 2006.
104. Kodali, P., Dhawan, N., Depci, T., Lin, C.L., Miller, D., Particle damage and exposure analysis in HPGR crushing of selected copper ores for column leaching. *Miner. Eng.*, 24, 1478–1487, 2011.
105. Ghorbani, Y., Becker, M., Peterson, J., Morar, S.H., Mainza, A., Franzidis, J.P., Use of X-ray computed tomography to investigate crack distribution and mineral dissemination in sphalerite ore particles. *Miner. Eng.*, 24, 12, 1249–1257, 2011.
106. Vizcarra, T.G., Wightman, E.M., Johnson, N.W., Manlapig, E.V., The effect of breakage mechanism on the mineral liberation properties of sulfide ores. *Miner. Eng.*, 23, 5, 374–382, 2010.
107. Genç, Ö. and Benzer, A.H., Effect of high-pressure grinding rolls (HPGR) pre-grinding and ball mill intermediate diaphragm grate design on grinding capacity of an industrial scale two-compartment cement ball mill classification circuit. *Miner. Eng.*, 92, 47–56, 2016.

108. Ghosh, S.N., *Cement and Concrete Science and Technology, volume I, Part I*, first ed, pp. 122–125, ISBN: ISBN: 81-85522-00-6, 1991.
109. Ballantyne, G.R., Hilden, M., van der Meer, F.P., Improved characterisation of ball milling energy requirements for HPGR products. *Miner. Eng.*, 116, 72–81, 2018.
110. Yin, W., Tang, Y., Ma, Y., Zuo, W., Yao, J., Comparison of sample properties and leaching characteristics of gold ore from jaw crusher and HPGR. *Miner. Eng.*, 111, 140–147, 2017.
111. Kelly, E.G. and Spottiswood, K., The breakage function: What is it really? *Miner. Eng.*, 405, 1990.
112. Patzelt, N., Knecht, H., Baum, W., Case made for high-pressure roll-grinding in gold plants. *Min. Eng.*, 47, 6, 524–529, 1995.
113. Lutz Klingmann, H., HPGR benefits at Golden Queen Soledad Mountain gold heap leaching project. *Randol Innovative Metallurgy Forum*, Perth, August 21–24, 2005.
114. Cleary, P.W. and Sinnott, M.D., Axial pressure distribution, flow behaviour and breakage within a HPGR investigation using DEM. *Miner. Eng.*, 163, 106769, 2021.
115. Ma, F., Tao, D., Tao, Y., Liu, S., An innovative flake graphite upgrading process based on HPGR, stirred grinding mill, and nanobubble column flotation. *Int. J. Min. Sci. Technol.*, 31, 6, 1063–1074, 2021.
116. Yin, W. and Chen, K., Effect of the particle size and microstructure characteristics of the sample from HPGR on column bioleaching of agglomerated copper ore. *Hydrometallurgy*, 200, 105563, 2021.
117. Chen, J.H., Tang, D., Zhong, S.P., Zhong, W., Li, B.Z., The influence of microcracks on copper extraction by bioleaching. *Hydrometallurgy*, 191, 105243, 2020a.
118. Chen, K.Q., Yin, W.Z., Rao, F., Wu, J., Zhu, Z.L., Tang, Y., Agglomeration of fine sized copper ore in heap leaching through geopolymerization process. *Miner. Eng.*, 159, 106649, 2020b.
119. Anticoi, H., Guasch, E., Ahmad Hamid, S., Oliva, J., Alfonso, P., Bascompta, M., Sanmiquel, L., Escobet, T., Escobet, A., Parcerisa, D., Peña-Pitarch, E., Argelaguet, R., De Felipe, J.J., An improved high-pressure roll crusher model for tungsten and tantalum ores. *Minerals*, 8, 11, 483, 2018. https://doi.org/10.3390/min8110483.
120. Anticoi, H., Guasch, E., Hamid, S.A., Oliva, J., Alfonso, P., Garcia-Valles, M., Bascompta, M., Sanmiquel, L., Escobet, T., Argelaguet, R., Escobet, A., De Felipe, J.J., Parcerisa, D., Peña-Pitarch, E., Breakage function for HPGR: Mineral and mechanical characterization of tantalum and tungsten ores. *Minerals.*, 8, 4, 170, 2018. https://doi.org/10.3390/min8040170.

121. Campos, T.M., Bueno, G., Tavares, L.M., Confined bed breakage of fine iron ore concentrates. *Minerals*, 10, 8, 666, 2020. https://doi.org/10.3390/min10080666.
122. Morley, C., HPGR-FAQ the users who uses HPGR? *SAIMM*, 110, 17–20, 2010.

2

Challenges Related to the Flotation Process of Complex Phosphate Ores

Nheta Willie* **and Hlahane Morake D.**

Department of Metallurgy, Faculty of Engineering and Built Environment, Mineral Processing and Technology Research Centre, University of Johannesburg, Auckland, Johannesburg, South Africa

Abstract

The high demand for phosphate in agriculture, pharmaceutical, and food industries has increased the exploitation rate of phosphate deposits, which are characterized by their low-grade and complex mineralogy. Many beneficiation techniques can be used to process phosphate ores, with froth flotation being the most used for finely disseminated complex ores. Although flotation is one of the most prevalent techniques for beneficiating phosphate ores, it is also one of the most difficult due to several factors. In this chapter, the main challenges facing the concentration of phosphate rock by froth flotation are discussed. Some of the factors affecting the phosphate ore flotation process are presented, including water quality, pH, flotation reagents, and ore mineralogy. Considering these challenges, insight into possible future research directions for froth flotation of complex phosphate ores is provided.

Keywords: Sedimentary phosphate, direct-reverse flotation, water quality, process water, phosphate rock

Abbreviations

PR	Phosphate rock
TSP	Triple superphosphate
MAP	Monoammonium phosphate

**Corresponding author*: wnheta@uj.ac.za

Elvis Fosso-Kankeu, Bhekie B. Mamba and Antoine F. Mulaba-Bafubiandi (eds.) Recovery of Values from Low-Grade and Complex Minerals: Development of Sustainable Processes, (41–64) © 2024 Scrivener Publishing LLC

DAP	Diammonium phosphate
SSP	Single superphosphate
P_2O_5	Phosphorus pentoxide
BPL	Bone phosphate of lime
XRD	X-ray powder diffraction
SEM	Scanning electron microscopy
CMC	Carboxymethyl cellulose
PAMS	poly-(acrylic acid-co maleic acid) sodium salt
PBTCA	2-phosphonobutane-1,2,4-tricarboxylic acid
EMS	Environmental management systems
TSS	Total suspended solids
TDS	Total dissolved solids
EC	Electrical conductivity

2.1 Introduction to the Geology of Complex Phosphate Ores

Phosphate rock (PR) is the principal natural source of phosphorus, which is important for the manufacture of fertilizers and phosphoric acid [1, 2]. Most of the global phosphate production is used in agriculture. However, the most important use is in the production of fertilizers because all fertilizers have N (nitrogen)–P (phosphorus)–K (potassium) grouping [3, 4]. PR is usually used to produce chemical fertilizers, such as triple superphosphate (TSP), monoammonium phosphate (MAP) or diammonium phosphate (DAP) and single superphosphate (SSP) [5]. Many other products are made from phosphate rock, including products for the food and pharmaceutical industries.

The mined and processed phosphate is usually referred to as phosphate rock and described as phosphorus pentoxide (P_2O_5) [6]. According to Gharabaghi *et al.*, 2010, cited in [7], in terms of phosphate ores, they are classified into three main classes determined by their origin and deposition. These are:

i. Sedimentary deposit of marine origin
ii. Igneous and metamorphic deposits
iii. Biogenic deposits from bird guano

Sedimentary phosphate deposits are found as layered seams of thickness ranging from less than one meter to several tens of meters [8–10]. They can be hundreds of kilometers long. Generally, they contain phosphorites (≥18% P_2O_5) or phosphate rocks (18% P_2O_5). While Ptáček [11] notes that sedimentary deposits contain several carbonate-fluorapatite varieties, they are grouped under the term "francolite". In addition, the prevalent gangue minerals found with sedimentary phosphate rocks are carbonates (calcite and dolomite), clay, and quartz [9].

Igneous deposits are formed from magma or lava, which resembles granitic lava [9, 10, 12]. They have limited texture or layering and are usually composed of black, white, and/or gray minerals. Pufahl and Groat [13] noted that magmatic phosphate deposits are commonly associated with carbonatites and alkaline intrusions with low silicon content. Phosphate rock concentrate from igneous origin is obtained from deposits mined mainly in Russia, Republic of South Africa, Brazil, Finland, and Zimbabwe [14]. Ores from igneous rocks are often of low grade (less than 5% P_2O_5) but can be processed into high quality products containing 35% to over 40% P_2O_5 [11].

Biogenic deposits are found mainly on oceanic islands. These are insoluble residues of calcium phosphate formed by the leaching process of ancient and/or fossil deposits of bird or bat excreta [15, 16].

Phosphate rock is produced in more than 40 countries [14]. The global production of PR is shown in Table 2.1 and the distribution of global phosphate resources is shown in Figure 2.1. According to Cisse and Mrabet [17], most of the phosphorus is produced from sedimentary and igneous deposits.

Sedimentary marine deposits are considered the most attractive type of phosphate resources and are expected to experience high demand in the global market. They also account for the largest market share. In South Africa, the commercially important phosphate deposits are proven to be found in magmatic and marine sedimentary rocks. The magmatic deposits are located in the Phalaborwa, Schiel and Spitskop complexes in the Limpopo Province [18]. The sedimentary deposits are derived from either offshore or onshore deposits and extend along the west coast of South Africa [19]. The ore consists of common phosphate minerals, mainly apatite and quartz. Apatite, as shown in Figure 2.2 [20], is a phosphate mineral and the most abundant phosphorus-bearing mineral. It has a general formula of $Ca_{10}(PO_4)6X_2$, where X is typically F (fluorapatite, FAp), OH (hydroxyapatite, OHAp), or Cl (chlorapatite, ClAp) [21].

Table 2.1 Global phosphate rock mine production in 2019 [14].

Country	Production (Mt)
China	95
US	23.3
Morocco	35.5
Russia	13.1
Jordan	9.2
Saudi Arabia	6.5
Egypt	5
Brazil	4.7
Vietnam	4.6
Tunisia	4.1
Peru	4
Senegal	3.4
Israel	2.8
Australia	2.7
South Africa	2.1
Syria	2
India	1.5
Kazakhstan	1.5
Algeria	1.3

2.2 Phosphate Rock Beneficiation Processes

The extraction of phosphate rock from mining operations follows a series of processing steps that include comminution (crushing, milling, grating, screening, and classification), mineral concentration processes (flotation, gravity, and magnetic separation), and sometimes chemical and thermal processes (acidification and calcination) [1, 14, 22]. Phosphate beneficiation

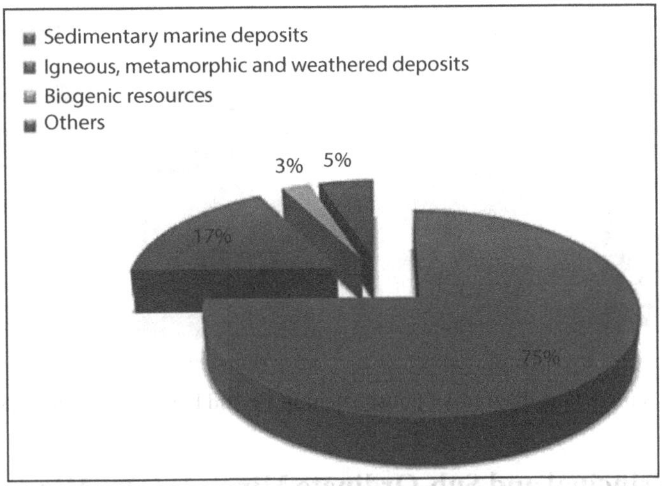

Figure 2.1 Distribution of the world's phosphate resources [11].

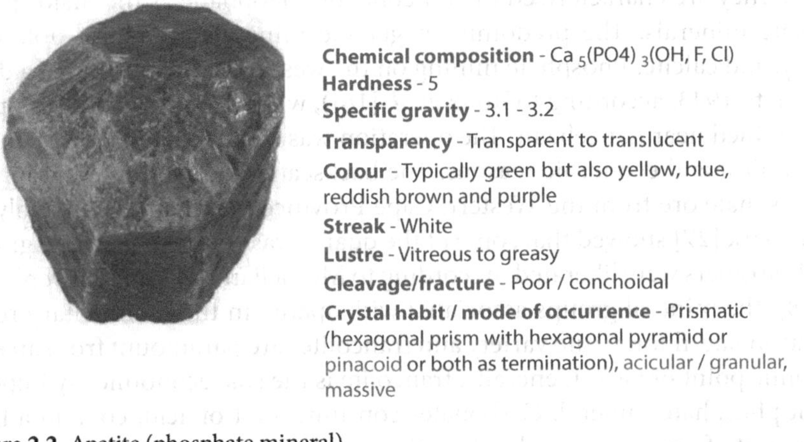

Figure 2.2 Apatite (phosphate mineral).

processes aim to refine the apatite mineral and remove gangue such as various types of silica, dolomite, and/or calcite. A saleable phosphate rock of normally +30% P_2O_5 or 65.5% bone phosphate of lime (BPL) is produced based on requirements of the downstream processes [23–25].

It is of utmost importance that the processing or beneficiation of phosphate rock rely on the geological origin. The mineralogy of the ore body is critical from a metallurgical standpoint. Figure 2.3 shows a simplified flow chart of phosphate mining and beneficiation processes.

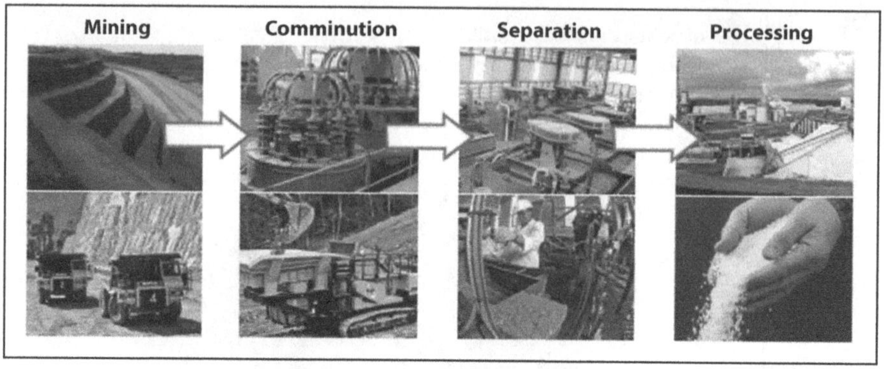

Figure 2.3 Simple flow diagram of phosphate mining and beneficiation process [26].

2.2.1 Principal and Sub-Ordinate Minerals (Ore Mineralogy)

Kawatra and Carlson [7] claim that most of the largest phosphate deposits are sedimentary deposits. They are found in Morocco, China, and Florida, US. They are characterized by francolite or collophane as the major phosphate minerals. The predominant gangue minerals are silica, dolomite, clay, and calcite. Phosphate mining on the west coast of South Africa dates back to 1943, according to Roux *et al.* [18]), when 16% P_2O_5 ore was open-pit mined near Langebaan. The operation was stopped after a few years for technical and economic reasons. The backscatter electron micrographs of phosphate ore from the Western Cape Province near Langebaan analyzed by Clark [27] showed that some of the quartz was trapped in apatite grains, while others were liberated. According to McClellan and Van Kauwenbergh [28], the mineral groups associated with apatite in the sedimentary rocks (carbonate-fluorapatite variety and francolite) are paramount from an economic point of view. Generally, francolite is the sole economically important phosphate mineral. Carbonates consume a lot of acid, contain a high amount of impurities, and reduce the grade of the ores [18, 29, 30].

The mineralogical process report of the sedimentary phosphate sample by Bastin [31] indicated four occurrences of the main mineral quartz and suggested the processing approach shown in Table 2.2.

According to Clark [27], sedimentary phosphate ores found in South Africa contain the mineralogy shown in Table 2.3. Minerals detected by X-ray powder diffraction (XRD) included quartz and apatite. Other minerals were observed by scanning electron microscopy (SEM) but present only in trace amounts. Clark reported that apatite was the predominant

Table 2.2 Quartz occurrence and possible beneficiation approach [31].

Quartz occurrence	Beneficiation approach
Free rounded grains, mainly in the coarser fraction	If reverse cationic flotation (Amine collector) was applied for this ore, these coarse free rounded grains would not be floated.
Unliberated grains cemented by a phosphatic phase	A finer grind would be required to liberate and separate these grains.
Grain coated by a phosphatic phase	These grains being coated by apatite behave like apatite in the flotation system, an attritioning operation within the flowsheet could clean these grains and improve flotation selectivity.
Angular clastic grains included in phosphatic grains	These fines quartz locked in the phosphatic elements are unavoidable. Their separation would probably be uneconomical.

P-bearing mineral, although crandallite, an Al-bearing apatite, was also present in trace amounts.

Lisiansky *et al.* [22] note that phosphate rock processing can often be simplified and made more economical by considering the distribution of P_2O_5 and impurities as a function of the ratio between the size fractions of mined and 'crushed' sediment phosphates. For this reason, comminution and size control are related on the goal of understanding the distribution of P_2O_5 and associated impurities. In mineral processing, classification is primarily used to separate particles by size. In phosphate rock processing, size control of classification is critical to achieve an effective enrichment ratio. Size control can determine the separation of apatite and silica.

According to Metso [32], there are two predominant size control methods, namely screening, which uses geometric patterns for size control, and classification, which uses particle motion for size control. In phosphate processing, size control is primarily achieved by classification because the mineral particles are too fine to be sorted by sieving. In some sedimentary phosphate mines, hydrocyclones and cross-flow separators are used for classification.

Table 2.3 Mineralogical composition of sedimentary phosphate ore [27].

Mineral	Formula	Relative abundance
Quartz	SiO_2	xxx
Apatite	$Ca_5(PO_4,CO_3)_3F$	xx
Crandallite	$CaAl_3(PO_4)_2(OH)_5\bullet(H_2O)$	✓
K-feldspar	$KAlSi_3O_8$	✓
Zircon	$ZrSiO_4$	✓
Rutile	TiO_2	✓
Pentlandite	$(Fe,Ni)_9S_8$	✓
Pyrite	FeS_2	✓
Fe-oxide/hydroxide	$FeOOH/Fe_2O_3$	✓
Clay (kaolinite)	$Al_2Si_2O_5(OH)_4$	✓
Monazite	(Ce, La, Nd, Th)PO_4	✓

xxx 20-50 mass%
xx 5-20 mass%
✓ observed by SEM

2.3 Froth Flotation of Sedimentary Phosphate Ore

The most effective beneficiation process for treating finely disseminated sedimentary phosphate ores is froth flotation. About 60% of the phosphate worldwide is produced through it [33–35]. It is very efficient in the removal of carbonate and silicate gangue minerals [36]. Froth flotation cannot be considered in isolation, but only in conjunction with other variables. However, it is well documented that flotation systems are complex, and they depend on several interrelated variables. Klimpel (1995), cited in [36], states that the flotation system consists of three main categories: (i) chemical components, which include collectors, depressants, frothers, pH adjusters, and other reagents required in flotation systems; (ii) equipment components, which include the type of flotation equipment, flotation cell size, flotation circuit configuration, and other parameters specific to particular flotation machines; and (iii) operating components, which include the mineralogy of the ore fed, particle size, pulp density, feed rate, and temperature. According to Hoang *et al.* [37], flotation is also affected by flotation cell hydrodynamics, surface properties of the ore and the gas bubbles.

Contact angle theory remains one of the top topics in flotation chemistry [38]. The contact angle simply describes the hydrophobic or hydrophilic nature of the solid at the solid–liquid–gas interface. Drelich and Marmur [39] argue that despite the tremendous progress in characterizing particles and their surfaces and understanding particle–bubble interactions, supported by detailed records of gas bubble attachment to both bulk samples and particles, the flotation process still correlates poorly with the wetting properties of particle surfaces. Contact angles, often used to describe mineral surface wettability, remain among the most controversial, misunderstood, and misinterpreted values in the mineral processing literature.

Studies have been conducted worldwide on the dosing and conditioning of reagents for phosphate processing, and many variables and procedures have been identified as critical for efficient flotation, either directly, inversely, or a combination of both [1, 40, 41]. Ruan *et al.* [1] stated that flotation reagents, such as depressants and collectors, play a major role on the selectivity of phosphate from gangue minerals. Depressants form a hydrophilic film at the surface of gangue minerals, prohibiting them from floating. Three types of froth flotation processes are commonly used for phosphate treatment, namely direct flotation and reverse flotation, or a combination of the two. The conventional "Crago" double flotation process is used in Florida, US for double removal of silica from phosphate ore [29, 42]. A new reverse "Crago" process using amine fatty acids was developed as a result of fatty acids, which are of poor quality, expensive, and meeting stringent environmental regulations [1].

- In direct flotation, the concentrate is floated and collected from the lip of a flotation cell or column. The desired main phosphate mineral (apatite) is floated and impurities such as silicates, carbonates and clays are separated [43–45].
- Reverse flotation is the opposite of direct flotation, in which the impurities are suspended and rejected by the lip of a flotation cell or column, while the main mineral of the phosphate (apatite) is separated as a concentrate [1, 44, 46].

2.3.1 Collectors Used in Phosphate Rock Flotation

Direct flotation of phosphate with fatty acid has been practiced in Florida, US since the 1920s. Arthur Crago patented the double flotation process using a 'negative ion agent' for coarser flotation followed by a 'positive ion agent' to remove silica [47]. Kawatra and Carlson [7] point out that there are generally four main categories of chemical reagent collectors used in phosphate flotation: nonionic, cationic, anionic, and amphoteric.

- *Non-ionic collectors* increase an already hydrophobic surface by coating at the surface e.g. fuel oil.
- *Cationic collectors* are amines consisting of positively charged cations.
- *Anionic collectors* are negatively charged carboxylate groups.
- *Amphoteric collectors* can either be cationic or anionic depending on the pH of the solution.

Table 2.4 shows various types of collectors commonly used in the flotation of phosphate ore. The importance of collector selection and water quality is inextricably linked. The flotation performance using amine or cationic collectors is affected by water quality.

The control of water quality is very difficult during fatty acid or anionic flotation processes. Inorganic acids in the water cause problems. Turbid water is detrimental to separation processes, and slightly alkaline water is preferable to reduce reagent costs [49]. Tall oil and oxidized petroleum are the main sources of anionic collectors for phosphate flotation [50]. However, they have the disadvantage of requiring operation at high temperatures, poor selectivity, and high reagent consumption [43]. Fatty acids modified by chlorination, hydroxylation, and etherification work better at room temperature [51]. Cationic collectors are used to remove silicates [52]. According to Vieira and Peres [53], ether diamine is more effective in the flotation of medium to coarse-grained quartz, while monoamines perform better in the flotation of fine fraction. Sodium dodecyl sulfonate and sulfosuccinate are used as auxiliary reagents due to their low frothiness [54]. Cationic collectors have the advantage of performing better at low temperatures and low reagent dosages. However, they are affected by slurries and are difficult to defoam during the flotation process. Therefore, they are good for coarse-grained ores and poor for fine-grained ores [1].

Amphoteric collectors are suitable for dolomite gangue removal [55]. They are effective over a broad pH range and are rarely affected by temperature and solutes. Higher recoveries of about 90% at 32% grade have been reported [1, 55]. According to Ruan *et al.* [1], amphoteric collectors can selectively remove calcite from apatite. However, amphoteric collectors have never been used in practice.

Table 2.4 Collectors used in phosphate ore flotation [48].

Mineral floated	Ionic property	Collector
Phosphate/apatite	Anionic	Oleic acid
		Fatty acids
		Disodium dodecyl phosphate
		Sodium alizarin sulfonate
		Sodium dodecyl sulfate
		Ethoxylated sulfonates and phosphates
		Ethoxylated sulfonates
		Hydroxamates
		Sulfosuccinate, sulfosuccinamate Berol 912L, 8305 Phospholan PE65
	Cationic	Armac T Amine
		Tallow amine acetate
	Amphoteric	Sodium Sarcosinate
		Atrac 1562, 1563, 2600, 922
Carbonates	Anionic	Fatty acids
		Sulfonated fatty acids
		Disodium dodecyl phosphate
		Phosphoric esters
		Ethoxylated phosphoric esters
	Amphoteric	Aminopripionic acid
		Carboxyethal imidazoline
Silica	Cationic	Amine
	Nanionic	Ocytlphenol ethoxylate
	Amphoteric	Aminopripionic acid

Recently, many researchers are investigating the use of mixed collectors for the flotation of phosphate ores, focusing mainly on ionic and nonionic collectors [1, 56]. Mixing anionic with nonionic collectors improves selectivity. According to Derqaoui *et al.* [57], alkyl hydroxamic acid performed better than the conventional fatty acid/heating oil solution. It also drastically improved the recovery of francolite [58].

2.3.2 Depressants Used in Phosphate Rock Flotation

In phosphate flotation, different types of depressants are used depending on the mineral to be floated. In direct flotation, silica/quartz is depressed, while apatite is floated. Sis and Chander [48] claim that the choice of depressant to be used depends on the type of collector used in the process. Table 2.5 shows some of the depressants used to depress gangue or sometimes to depress some minerals.

Phosphate minerals (apatite) can be depressed by floating at a pH below 4.5 in anionic flotation [59]. At this pH, apatite is depressed by the presence of $CaH_2PO_4^+$ and $H_2PO_4^-$ ions. According to Zheng *et al.* [60], calcium ions bind with $H_2PO_4^-$ on the surface of apatite, leading to a decrease in active cites in an acidic media. Various inorganic acids, sulfate or oxalate salts, and phosphate salts can increase the depression of apatite at pH below 4 [61]. Sulfuric acid and its salts cannot be used to depress phosphate minerals

Table 2.5 Depressants used in phosphate ore flotation [48].

Depressed mineral	Depressant
Phosphate minerals	Aluminum sulfate and sodium
	Sodium carbonate
	Phosphoric acid
	Starch
Carbonates	Sodium Silicate
	Hydrofluoric acid
	Polysaccharides
Silica	Sodium silicate

because of the formation of gypsum. Therefore, phosphoric acid is used in the processing plant in the anionic reverse flotation of phosphate minerals.

In cationic reverse flotation, several depressants were investigated by Zhang and Snow [62]. Starch proved to be a better depressant for the removal of silica from phosphate at a finer size (-35 mesh). However, sodium tripolyphosphate performed better at coarser sizes (+35 mesh). Nagaraj *et al.* [63] demonstrated that polyacrylamide-based polymers have better selectivity in the separation of apatite from siliceous gangue.

Gangue minerals containing carbonate are removed by reverse flotation in the presence of a phosphate depressant [61, 64]. However, for calcareous ores containing silicate and carbonate gangue, a double float is employed. Carboxymethyl cellulose (CMC), citric acid, and naphthyl anthryl sulfonates have been reported to be effective depressants for dolomite [64]. Recently, eco-friendly depressants such as poly-(acrylic acid-co maleic acid) sodium salt (PAMS) [65] and 2-phosphonobutane-1,2,4-tricarboxylic acid (PBTCA) [66] for dolomite have been investigated, and they have proved to be strongly selective. Depression of dolomite by bacteria (*Bacillus subtilis* and *Mycobacterium phlei*) has been reported by Zheng *et al.* [60]. However, they depress both apatite and dolomite and there is need for further investigations.

Silicate minerals are effectively depressed by sodium silicate using fatty acids as a collector in direct flotation [67]. $NaHCO_3$ is an effective depressant of quartz during amine collector cationic flotation and during anionic flotation in the presence of sodium oleate and Ca^{2+} [68, 69]. According to Guimarães *et al.* [70], starch can depress quartz, but it has poor selectivity. However, it is widely used in Brazil for beneficiation of igneous phosphate ores.

2.3.3 Frothers Used in Phosphate Rock Flotation

The frothing and dispersant agents listed in Table 2.6 are generally used in the flotation of phosphate rock. The quality of the froth is determined by the type and concentration of the frothing agent used. Sodium silicate (Na_2SiO_3) is the most commonly used frothing agent for dispersion of silicate minerals [46].

2.3.4 Effect of pH on Flotation of Phosphate Ores

Potency of collectors depends on pH [6]. Alsafasfeh and Alagha [45] stated that different techniques, such as zeta potential, contact angle measurements,

Table 2.6 Some of the frothers and dispersants used in phosphate ores.

Frother	**Methyl Isobutyl Carbinol (MIBC)**
Dispersants	Sodium Silicate (Na_2SiO_3)

and quartz crystal microbalance with dissipation, were used to evaluate the effects of pH. These techniques can provide both qualitative and quantitative information on the change in surface properties of mineral particles due to a change in pulp pH [71]. The pH can improve the adhesion of collectors to the mineral surface. The surface chemistry of minerals can be altered by changing pH due to the change in surface charges. Most minerals are positively charged in acidic conditions and negatively charged in alkaline conditions according to Kawatra *et al.* (2001), cited in [72].

2.3.5 Equipment Used in Phosphate Rock Flotation

The equipment components in terms of the type of flotation cells for most phosphate processing are either a column, a conventional system, or a hybrid system consisting of the two flotation cells, depending on their reaction and performance. Froth flotation is an extremely versatile process for the physical separation of particles due to the availability of equipment to treat even finely disseminated ores [7].

Damodaran and Fahey [73] stated that mixing is important in the conditioning of phosphate with anionic reagents. The studies found that too little mixing, in terms of time or energy, can result in poor flotation performance. Excessive mixing, either in terms of time or energy, can affect the binding of the reagents and the resulting flotation performance.

2.4 Challenges Facing Flotation of Phosphate Rock

2.4.1 Water Quality

The problem of water quality is a long-standing and deeply rooted problem in the mining industry. Conservation and management of freshwater resources is one of the greatest challenges facing humanity in this century. Mining and mineral processing industries use water in all their activities, from exploration to mining, processing, and closure [74, 75].

Water quality has been recognized in previous studies as essential and critical to flotation performance [76]. Reuse of process water is an economic imperative in many operations. However, internal reuse is susceptible to problems associated with processing, such as ore dissolution and reagent addition, which later lead to water quality issues as water is reused in processing facilities.

It is an undeniable fact that one of the paramount challenges in the mineral processing and mining sector is their environmental management systems (EMS) and their contribution to the global transition to sustainable development. For this reason, water chemistry and quality in mineral processing is a non-negotiable imperative. There are studies that have been conducted to improve the understanding of water quality variations in flotation [76–78]. However, it is important to note that the studies are usually conducted on a case-by-case basis. Liu *et al.* [79] state that there is no framework to summarize existing studies, leading to poor identification of research gaps and directions of future research. The research study on ultrafine phosphate processing suggested that flotation recovery of P_2O_5 is inversely proportional to the hardness of the water that is used in grinding, conditioning, and flotation, and further stated that elevated levels of water hardness result in more silica reporting to the phosphate concentrate.

Groundwater seepage is among the possible water resources in the sedimentary phosphate mining [80]. The water quality of the groundwater is also open for possible deterioration by means of natural effects or due to unforeseeable operational consequences. There are significantly positive social implications when the use of fresh water from local suppliers is reduced and diverted to the communities. It is thus critical that its chemistry and influence on downstream processing are understood because the requirement for water is predicted to grow considerably in the next decade for both mining operations and for the surrounding communities.

Water quality is affected by many parameters, and they include, among others, total suspended solids (TSS), total dissolved solids (TDS), electrical conductivity (EC), iron (Fe) content, and typical contaminants such as, among others, magnesium (Mg), sulphate (SO_4), calcium (Ca), potassium (K), and sodium (Na) [81, 82]. Each of these parameters affect water quality either separately or combined. Accumulation of organic or inorganic species, suspended solids (SS), residual reagents, and bacteria negatively affect flotation performance although some positive effects have been reported [83].

The effects of Ca $(NO_3)_2$, KNO_3, KF, and KH_2PO_4 salts on the zeta potential of natural apatite were studied by Somasundaram and Zhang (1999), cited in [7]. It was found that natural apatite has an isoelectric point (IEP) of 5.9 in water. It is not affected by K^+ and NO^{3-} ions since they are not potential determining ions. However, IEP changed to 2.5 in the presence of Ca^{+2}, F^-, and the various PO_4 complexes (PO_4^{-3}, HPO_4^{-2}, and $H_2PO_4^-$), and zeta potential was affected over a wide range of pH. The studies further showed that the adsorption of Ca^{+2} ions on quartz at a basic pH can be detrimental to flotation processes because of the heterocoagulation of calcium carboxylate precipitates on the quartz surface.

Castro and Laskowski [84] found that the flotation of phosphate ores with fatty acids (anionic collector) requires the removal of Ca^{2+} and Mg^{2+} ions if seawater was used in flotation. Furthermore, Laskowski and Castro [85] stated that there is need to remove Mg^{2+} and Ca^{2+} ions during flotation of phosphate ores using fatty acids since the flotation efficiency is reduced due to the formation of precipitates of Ca/Mg fatty acid salts. The study by Wei *et al.* [86] revealed that water conductivity and froth stability are affected by flotation reagents. Frothers react with salt ions to form a more stable froth.

The influence of water quality in the phosphate beneficiation processes aimed at upgrading apatite mineral continues to be a challenge especially since many operations have their own specific limitations. The need for alternative processing methods that would require minimal usage of raw (municipal) water is a non-negotiable business imperative, and mining industries have to forge a collective pact of processing norms and standards of practice primarily driven for alternative methods to minimize and manage current water and energy footprints.

2.4.2 Mineralogy of the Phosphate Rock

Mineralogy plays a pivotal role in the flotation of PR. Different types of phosphate rocks must be subjected to different types of flotation processes depending on the characteristics of the ore [1]. Some of the phosphate minerals and their impact are shown in Table 2.7.

2.4.3 Particle Size Distribution Challenges

In phosphate flotation, like in many other mineral flotations such as sulphides, the fine and ultrafine particles are problematic compared to other intermediate size fractions. Generally, the flotation recovery and efficiency are impacted when fine material is treated. For finely disseminated phosphate ores, a high degree of mineral liberation increases creation of fine and ultrafine materials. These fines negatively affect the flotation efficiency

Table 2.7 Some of the phosphate elements and their impact on processing [87].

Symbol %	Impact
P_2O_5	Normally, a preferred rock is the one that yields P_2O_5 as phosphoric acid at lowest cost, and primary consideration is usually the delivered price of the rock per unit of P_2O_5.
SiO_2	The silica found in significant quantities leads to greater wear of the grinding apparatus and erodes the agitators and the pumps that handle slurries of attack. It can also negatively affect the filtration of gypsum in the production of phosphoric acid.
Al_2O_3 and Fe_2O_3	Alumina is undesirable because after dissolution of phosphate by the acid, they form insoluble phosphates that decrease water soluble P_2O_5 content in the acid product. In addition, they reduce the filterability of gypsum.
MgO	Magnesium in phosphoric acid has a double effect on the crystallization of gypsum. Combined with fluorine, it enhances crystallization, but over a certain limit (2%–3%), it increases the viscosity of the acid, which generates a rapid deterioration of the filtration rate.
Na_2O and K_2O	The Na_2O and K_2O contents are determined by flame spectrophotometry. This method is applicable to inorganic phosphates whose contents in Na_2O and K_2O are respectively less than 2.7% and 0.24%. Na and K react with fluorine and silica to form fluosilicates (Na_2SiF6 and K_2SiF_6), which can produce muds in the equipment and whose elimination can raise important problems.
CaO	Calcite increases the consumption of sulfuric acid for the attack. A relatively pure acid can be produced from ore containing large quantities of calcite. Calcite causes the formation of foam during the attack, especially nitric. This requires an increased amount of anti-foam, but they improve the reactivity of the ore, particularly appreciated in the production of phosphate fertilizers.

[1, 71, 88]. Hoang *et al.* state that fine and ultrafine materials reduce the bubble–particle attachment efficiency as a result of poor flotation kinetics and gangue entrainment.

Fine and coarse particles are affected differently by reagent concentration variations [88]. Generally, coarser particles are more sensitive to reagent concentration changes, thereby producing a better product. Finer particles have less sensitivity to reagent concentration changes explaining why it is difficult to process fine materials. According to King [89], the flotation principles are well understood and detailed in literature but building quantitative predictive models has proved to be very difficult due to many microprocesses involved.

2.5 Future Research Directions

Future research on processing of PR should focus on mineral liberation of finely disseminated complex ores. Many medium- to coarse-grained ores are becoming depleted, and many companies will now focus on fine-grained ores, ensuring the development of more selective and less toxic reagents that will impact less on the environment. For example, phosphoric acid and sulphuric acid and their derivatives have a detrimental effects to the environment. Impact of water recirculation on PR flotation process must be thoroughly investigated focusing on the accumulation of detrimental ions in water, as well as their removal from recirculated water. Phosphate waste valorization is another aspect with many industrial opportunities.

2.6 Conclusion

Flotation is the main beneficiation process of finely disseminated phosphate ores. It is, however, affected by many factors such as the mineralogical properties of the ore, water quality, and flotation reagents used. Coarse-grained ores are easy to process compared to finely grained ores. Fatty acids are the most used anionic collectors, and cationic amines are more common for the removal of silica. Sodium silicate is popular for depressing silica and organic reagents such as CMC, and naphthyl sulphonates are good for dolomite depression. The influence of process water and its complexities could not be generalized, and there has been progress in understanding the underlying tenets by means of predictive models for simulation packages. Some of the phosphate deposits are found in semi-arid and arid areas, and there is a need to come up with alternative sources of water supply. The potential use of alternative water stream in the downstream flotation process for the sedimentary phosphate is a possibility that could significantly contribute to water saving and sustainability. However, it is paramount to understand the impact of the quality of the alternative water on the downstream processes such as flotation circuits.

References

1. Ruan, Y., He, D., Chi, R., Review on beneficiation techniques and reagents used for phosphate ores. *Minerals*, 9, 4, 253, 2019.
2. Dolan, M., Phosphate rock 101, September 28, 2016. [Online]. Available: https://www.nexant.com/resources/phosphate-rock-101.
3. Bindraban, P., Dimkpa, C., Nagarajan, L., Roy, A., Rabbinge, R., Revisiting fertilisers and fertilisation strategies for improved nutrient uptake by plants. *Biol. Fertil. Soils*, 51, 8, 897–911, 2015.
4. Macura, B., Johannesdottir, S., Piniewski, M., Haddaway, N., Kvarnström, E., Effectiveness of ecotechnologies for recovery of nitrogen and phosphorus from anaerobic digestate and effectiveness of the recovery products as fertilisers: A system review protocol. *Environ. Evid.*, 8, 1, 1–9, 2019.
5. Elias, E., Teklu, B., Tefera, T., Effect of blended fertiliser application on bread wheat yield, agronomic efficiency and profitability on Nitisols of Southern Ethiopia. *S. Afr. J. Plant Soil*, 37, 4, 292–299, 2020.
6. Farid, Z., Abdennouri, M., Barka, N., Jannani, Y., Sadiq, M., Study of the effect of pH, conditioning and flotation time on the flotation efficiency of phosphate ores by a soybean oil collector. *J. Met. Mater. Miner.*, 32, 1, 101–108, 2022.
7. Kawatra, S. and Carlson, J., *Beneficiation of Phosphate Ore*, 2013.
8. Simandl, G., Paradis, S., Fajber, R., *Sedimentary phosphate deposits mineral deposit profile F07. Geological Fieldwork*, pp. 217–222, 2012.
9. El Bamiki, R., Raji, O., Ouabid, M., Elghali, A., Khadiri Yazami, O., Bodinier, J., Phosphate rocks: A review of sedimentary and igneous occurrences in Morocco. *Minerals*, 11, 10, 1137, 2021.
10. Hellal, F., El-Sayed, S., Zewainy, R., Amer, A., Importance of phosphate pock application for sustaining agricultural production in Egypt. *Bull. Natl. Res. Centr.*, 43, 1–11, 2019.
11. Ptáček, P., Apatites and their synthetic analogues, in: *Synthesis, Structure, Properties and Applications*, I.D. Team (Ed.), p. 335, InTech, Rijeka, Croatia, 2016.
12. He, G. and Zhou, Y., Geology of phosphate rock in China: distribution, rock type and metallogenic perspective. *Glob. Environ. Res.*, 19, 1, 91–96, 2015.
13. Pufahl, P. and Groat, L., Sedimentary and igneous phosphate deposits: formation and exploration: An invited paper. *Econ. Geol.*, 112, 3, 483–516, 2017.

14. Amar, H., Benzaazoua, M., Elghali, A., Hakkou, R., Taha, Y., Waste rock reprocessing to enhance the sustainability of phosphate reserves: A critical review. *J. Clean. Prod.*, 135151, 2022.
15. Harraz, H., *Topic 10: Phosphite Ore Deposits*, 2016, [Online]. Available: https://www.researchgate.net/profile/Hassan_Harraz/publication/301862850_PHOSPHITE_ORE_DEPOSITS/links/572a424c08aef5d48d30c9d5.pdf. [Accessed 19 August 2020].
16. Auer, G., Reuter, M., Hauzenberger, C., Piller, W., The impact of transport processes on rare earth element patterns in marine authigenic and biogenic phosphates. *Geochim. Cosmochim. Acta*, 203, 140–156, 2017.
17. Cisse, L. and Mrabet, T., World phosphate production: Overview and prospects. *Phosphorus Res. Bull.*, 15, 21–25, 2004.
18. Roux, E., De Jager, D., Du Plooy, J., Nicotra, A., Van der Linde, G., De Waal, P., Phosphate in South Africa. *J. South. Afr. Inst. Min. Metall.*, 89, 5, 129–139, 1989.
19. Louw, I., Potential radiological impact of the phosphate industry in South Africa on the public and the environment (Paper 1). *J. Environment. Radioact.*, 217, 106214, 2020.
20. Auckland, U. o., Geology rocks and minerals, 2005. [Online]. Available: https://flexiblelearning.auckland.ac.nz/rocks_minerals/minerals/apatite.html. [Accessed 12 July 2020].
21. Elliott, J., Wilson, R., Dowker, S., Apatite structures, in: *Advances in X-ray Analysis*, vol. 45, pp. 172–181, 2002.
22. Lisiansky, L., Baker, M., Larmour-Ship, K., Elyash, O., A tailor made approach for the beneficiation of phosphate rock, in: *Beneficiation of Phosphates*, pp. 55–61, 2016.
23. Kawatra, S. and Carlson, J., *Beneficiation of Phosphate Ore*, Society for Mining, Metallurgy, and Exploration, 2013.
24. Aleksandrova, T., Elbendari, A., Nikolaeva, N., Beneficiation of a low-grade phosphate ore using a reverse flotation technique. *Miner. Process. Extr. Metall. Rev.*, 43, 1, 22–27, 2022.
25. Aarab, I., Derqaoui, M., Abidi, A., Yaacoubi, A., El Amari, K., Etahiri, A., Baçaoui, A., Direct flotation of low-grade Moroccan phosphate ores: A preliminary micro-flotation study to develop new beneficiation routes. *Arab. J. Geosci.*, 13, 1–11, 2020.
26. Foskor, Foskor recycles water to preserve the environment, 2005.
27. Clark, W., *Mineralogical Analysis of Core and Sized Composite Samples From The Elandsfontein Phosphate Project*, Mintek, Internal Report, Unpublished, Randburg, South Africa, 2014.
28. McClellan, G. and Van Kauwenbergh, S., Mineralogy of sedimentary apatites. *Geological Society, London, Special Publications*, 52, 1, 23–31, 1990.

29. Zhang, P., Zheng, S., Song, W., Feng, C., Moudgil, B., Xiao, W., Zhang, D., "Promising options for solving the dolomite problem of the Florida phosphate resources—a brief review,". *Min. Metall. Explor.*, 36, 117–129, 2019.
30. Abbes, N., Bilal, E., Hermann, L., Steiner, G., Haneklaus, N., Thermal beneficiation of Sra Ouertane (Tunisia) low-grade phosphate rock. *Minerals*, 10, 11, 937, 2020.
31. Bastin, D., *Process Mineralogy Report*, Prayton Technologies SA, University of Lieg, 2015.
32. Metso, Basics in mineral processing, 2018. [Online]. Available: https://www.metso.com/contentassets/0efc5d1a7c5a4357baecc5e990dc1fe7/basics-in-mineral-processing-handbook-18-lr.pdf. [Accessed 17 November 2018].
33. Abouzeid, A., Physical and thermal treatment of phosphate ore - an overview. *Int. J. Miner. Process.*, 85, 4, 59–84, 2008.
34. Yang, X., Tamm, K., Piir, I., Kuusik, R., Trikkel, A., Tonsuaadu, K., Evaluation of Estonian phosphate rock by flotation. *Miner. Eng.*, 171, 107127, 2021.
35. Mohammadkhani, M., Noaparast, M., Shafaei, S., Amini, A., Amini, E., Abdollahi, H., Double reverse flotation of a very low grade sedimentary phosphate rock, rich in carbonate and silicate. *Int. J. Miner. Process.*, 100, 3-4, 157–165, 2011.
36. Kawatra, S.K. and Carlson, J., *Beneficication of Phosphate Ore*, Society of Mining, Metallurgy & Exploration (SME, Colorado, USA, 2014.
37. Hoang, D., Hassanzadeh, A., Peuker, U., Rudolph, M., Impact of flotation hydrodynamics on the optimisation of fine-grained carbonaceous sedimentary apatite ore beneficiation. *Powder Technol.*, 345, 223–233, 2019.
38. Drelich, J., Boinovich, L., Chibowski, E., Della Volpe, C., Hołysz, L., Marmur, A., Siboni, S., Contact angles: History of over 200 years of open questions. *Surf. Innov.*, 8, 1–2, 3–27, 2019.
39. Drelich, J. and Marmur, A., Meaningful contact angles in flotation systems: Critical analysis and recommendations. *Surf. Innov.*, 6, 1–2, 19–30, 2017.
40. Derqaoui, M., Aarab, I., Abidi, A., Yaacoubi, A., El Amari, K., Etahiri, A., Baçaoui, A., Review of the reagents used in the direct flotation of phosphate ores. *Arab. J. Geosci.*, 15, 1–21, 2022.
41. Yang, X., Tamm, K., Piir, I., Kuusik, R., Trikkel, A., Tonsuaadu, K., Evaluation of Estonian phosphate rock by flotation. *Miner. Eng.*, 171, 107127, 2021.
42. Guan, C., Removal of coarse fraction from phosphatic clays for clay settling area reduction and additional phosphate recovery, in: *Beneficiation of Phosphates: Sustainability, Critical Materials, Smart Processes*, p. 28, 2019.
43. Ruan, Y., He, D., Chi, R., Review on beneficiation techniques and reagents used for phosphate ores. *Minerals*, 9, 4, 253, 2019.
44. Liu, X., Zhang, Y., Liu, T., Cai, Z., Sun, K., Characterization and separation studies of a fine sedimentary phosphate ore slime. *Minerals*, 7, 6, 94, 2017.

45. Alsafasfeh, A. and Alagha, L., Recovery of phosphate minerals from plant tailings using direct froth flotation. *Minerals*, 7, 8, 145, 2017.
46. Liu, X., Li, C., Luo, H., Cheng, R., Liu, F., Selective reverse flotation of apatite from dolomite in collophanite ore using saponified gutter oil fatty acid as a collector. *Int. J. Miner. Process.*, 165, 20–27, 2017.
47. Dopico, P., Lipowsky, G., Mbonambi, M., Mahlangu, L., Makin, B., Pedain, K., Silva, W., *Phosphate Beneficiation with Novel Collectors for Direct and Reverse Flotation: Beyond Low Cost Fatty Acids*, Society for Mining, Metallurgy & Exploration, Englewood, CO, USA, 2015.
48. Sis, H. and Chander, S., Reagents used in the flotation of phosphate ores: A critical review. *Miner. Eng.*, 16, 7, 577–585, 2003.
49. Reta, G., Dong, X., Li, Z., Su, B., Hu, X., Bo, H., Yu, D., Wan, H., Liu, J., Li, Y., Xu, G., Environmental impact of phosphate mining and beneficiation: Review. *Int. J. Hydrol.*, 2, 4, 424–431, 2018.
50. Ramezani, M., Lashkarbolooki, M., Abedini, R., Experimental investigation of different characteristics of crude oil on the interfacial activity of anionic, cationic and nonionic surfactants mixtures. *J. Pet. Sci. Eng.*, 214, 110485, 2022.
51. Huang, Q., Huang, J., Zhou, H., Pan, Z., Ping, X., Synthesis and application of a flotation collector for collophanite, in: *Beneficiation of Phosphates: New Thought, New Technology, New Development*, Zhang, P., Miller, J., El-Shall, H. (Eds.), pp. 359–364, 2012.
52. Amirech, A., Bouhenguel, M., Kouachi, S., Two-stage reverse flotation process for removal of carbonates and silicates from phosphate ore using anionic and cationic collectors. *Arab. J. Geosci.*, 11, 19, 593, 2018.
53. Vieira, A. and Peres, A., The effect of amine type, pH, and size range in the flotation of quartz. *Miner. Eng.*, 20, 10, 1008–1013, 2007.
54. Oliveira, M., Santana, R., Ataíde, C., Barrozo, M., Recovery of apatite from flotation tailings. *Sep. Purif. Technol.*, 79, 1, 79–84, 2011.
55. Li, J., Nie, G., Li, J., Zhu, Z., Wang, Z., Flotation separation of quartz and dolomite from collophane using sodium N-dodecyl-β-amino propionate and its adsorption mechanism. *Colloids Surf. A Physicochem. Eng. Asp.*, 641, 128586, 2022.
56. Tang, Y., Yin, W., Kelebek, S., Magnesite-dolomite separation using potassium cetyl phosphate as a novel flotation collector and related surface chemistry. *Appl. Surf. Sci.*, 508, 145191, 2020.
57. Derqaoui, M., Aarab, I., Abidi, A., Yaacoubi, A., El Amari, K., Etahiri, A., Baçaoui, A., Review of the reagents used in the direct flotation of phosphate ores. *Arab. J. Geosci.*, 15, 1–21, 2022.

58. Wang, X., Nguyen, A., Miller, J., Selective attachment and spreading of hydroxamic acid–alcohol collector mixtures in phosphate flotation. *Int. J. Miner. Process.*, 78, 2, 122–130, 2006.
59. Ruan, Y., Zhang, Z., Luo, H., Xiao, C., Zhou, F., Chi, R., Effects of metal ions on the flotation of apatite, dolomite and quartz. *Minerals*, 8, 4, 141, 2018.
60. Zheng, X., Arps, P., Smith, R., Adhesion of two bacteria onto dolomite and apatite: their effect on dolomite depression in anionic flotation. *Int. J. Miner. Process.*, 62, 1-4, 159–172, 2001.
61. Chen, Y., Feng, Q., Zhang, G., Liu, D., Liu, R., Effect of sodium pyrophosphate on the reverse flotation of dolomite from apatite. *Minerals*, 8, 7, 278, 2018.
62. Zhang, P. and Snow, R., Evaluation of phosphate depressants in the phosphate/silica system. *Min. Metall. Explor.*, 26, 101–104, 2009.
63. Nagaraj, D., Rothenberg, A., Lipp, D., Panzer, H., Low molecular weight polyacrylamide-based polymers as modifiers in phosphate beneficiation. *Int. J. Miner. Process.*, 20, 3-4, 291–308, 1987.
64. Zheng, X. and Smith, R., Dolomite depressants in the flotation of apatite and collophane from dolomite. *Miner. Eng.*, 10, 5, 537–545, 1997.
65. Yang, B., Zhu, Z., Sun, H., Yin, W., Hong, J., Cao, S., Tang, Y., Zhao, C., Yao, J., Improving flotation separation of apatite from dolomite using PAMS as a novel eco-friendly depressant. *Miner. Eng.*, 156, 106492, 2020.
66. Yang, B., Zhu, Z., Yin, W., Sun, Q., Sun, H., Han, H., Sheng, Q., Yao, J., Selective adsorption of an eco-friendly and efficient depressant PBTCA onto dolomite for effective flotation of fluorapatite from dolomite. *Chem. Eng. J.*, 400, 125780, 2020.
67. Dho, H. and Iwasaki, I., Role of sodium silicate in phosphate flotation. *Miner. Metall. Process.*, 7, 4, 215–221, 1990.
68. Sayilgan, A. and Arol, A., Effect of carbonate alkalinity on flotation behavior of quartz. *Int. J. Miner. Process.*, 74, 1-4, 233–238, 2004.
69. El-Mofty, S. and El-Midany, A., Role of calcium ions and their interaction with depressants in phosphate flotation. *Chem. Pap.*, 72, 2641–2646, 2018.
70. Guimarães, R., Araujo, A., Peres, A., Reagents in igneous phosphate ores flotation. *Miner. Eng.*, 18, 2, 199–204, 2005.
71. Derhy, M., Taha, Y., Hakkou, R., Benzaazoua, M., Review of the main factors affecting the flotation of phosphate ores. *Minerals*, 10, 12, 1109, 2020.
72. Sheni, N., Corin, K., Wiese, J., Considering the effect of pulp chemistry during flotation on froth stability. *Miner. Eng.*, 116, 15–23, 2018.
73. Damodaran, R. and Fahey, M., Effect of conditioning variables on impeller power in anionic conditioning of phosphate slurries. *Sep. Sci. Technol.*, 30, 11, 2385–2402, 1995.

74. Muzinda, I. and Schreithofer, N., Water quality effects on flotation: Impacts and control of residual xanthates. *Miner. Eng.*, 125, 34–41, 2018.
75. Li, Y., Xie, S., Zhao, Y., Xia, L., Li, H., Song, S., The life cycle of water used in flotation: A review. *Min. Metall. Explor.*, 36, 385–397, 2019.
76. Muzenda, E., An investigation into the effect of water quality on flotation performance. *World Acad. Science, Eng. Technol. Int. J. Chem. Mol. Eng.*, 237–241, 2010.
77. Teague, A. and Lollback, M., The beneficiation of ultrafine phosphate. *Miner. Eng.*, 27, 52–59, 2012.
78. Boujounoui, K., Abidi, A., Bacaoui, A., El Amari, K., Yaacoubi, A., The influence of water quality on the flotation performance of complex sulphide ores: Case study at Hajar Mine, Morocco. *J. South. Afr. Inst. Min. Metall.*, 115, 12, 1243–1251, 2015.
79. Liu, W., Moran, C., Vink, S., A review of the effect of water quality on flotation. *Miner. Eng.*, 53, 91–100, 2013.
80. Hlahane, M., Nheta, W., Makhatha, M., *The Impact of Using Mining Pit Water on Flotation Performance of A Sedimentary Phosphate Ore*, 2019.
81. Uddin, M., Nash, S., Olbert, A., A review of water quality index models and their use for assessing surface water quality. *Ecol. Indic.*, 122, 107218, 2021.
82. Chen, C., Wang, S., Kim, H., Pan, S., Fan, C., Lin, Y., Non-conventional water reuse in agriculture: A circular water economy. *Water Res.*, 199, 117193, 2021.
83. Le, T., Schreithofer, N., Dahl, O., Dissolution test protocol for estimating water quality changes in mineral processing plant operating with closed water circulation. *Minerals*, 10, 8, 653, 2020.
84. Castro, S. and Laskowski, J., Froth flotation in saline water. *KONA Powder Part. J.*, 29, 4–15, 2011.
85. Laskowski, J. and Castro, S., Flotation in concentrated electrolyte solutions. *Int. J. Miner. Process.*, 144, 50–55, 2015.
86. Wei, T., Peng, Y., Vink, S., The joint action of saline water and flotation reagents in stabilizing froth in coal flotation. *Int. J. Miner. Process.*, 148, 15–22, 2016.
87. Chaabouni, A., Chtara, C., Nzihou, A., Elfeki, H., Study the nature and the effects of the impurities of phosphate rock in the plants of production of phosphoric acid. *J. Adv. Chem.*, 7, 2, 1296–1299, 2014.
88. Santana, R., Duarte, C., Ataíde, C., Barrozo, M., Flotation selectivity of phosphate ore: Effects of particle size and reagent concentration. *Sep. Sci. Technol.*, 46, 9, 1511–1518, 2011.
89. King, R., *Modeling and Simulation of Mineral Processing Systems*, Elsevier, 2001.

3

Increasing Ionic Strength and Oxyhydroxo Species in Process Water on the Floatability of Chalcopyrite and Pentlandite for a Selected Cu–Ni Bearing Ore Flotation

Malibongwe S. Manono[1]*, Katlego M. Matibidi[2], Lisa L. October[1], Catherine K. Thubakgale[2] and Kirsten C. Corin[1]

[1]*Centre for Minerals Research, Department of Chemical Engineering, University of Cape Town, Rondebosch, South Africa*
[2]*Department of Chemical & Metallurgical Engineering, Vaal University of Technology, Vanderbijlpark, South Africa*

Abstract
For mining to be resilient against process water quality changes, researchers need to consider process water chemistry as an important variable in predicting flotation performance. Therefore, this study investigates the interactive effect of plant water ionic strength and pulp pH during the flotation of a Cu–Ni bearing ore. The ionic strength (I) of plant water was varied between 0.0242 mol.dm^{-3} and 0.242 mol.dm^{-3}; the pH was adjusted to pH 9 and pH 11. Water recoveries indicate that the stability of the froth increased in the presence of hydroxyl ions, known to be present at more alkaline pH. At pH 9, solids recoveries increased with increasing I. However, at pH 11, the opposite trend was observed. The decrease in solids recoveries with increasing I at pH 11 can be attributed to the depressive nature of oxyhydroxo species. Cu and Ni grades decreased with increasing I at pH 9. Whereas at pH 11, Cu and Ni grades increased with increasing I, suggesting an improvement in gangue depression at higher I and pH. These results show that water in flotation may have effects at both the air–water interface, as can be inferred from observations on the stability of the froth, as well as the air–mineral and mineral–water

**Corresponding author*: malibongwe.manono@uct.ac.za

Elvis Fosso-Kankeu, Bhekie B. Mamba and Antoine F. Mulaba-Bafubiandi (eds.) Recovery of Values from Low-Grade and Complex Minerals: Development of Sustainable Processes, (65–86) © 2024 Scrivener Publishing LLC

interfaces, as can be inferred from total solids, Cu, and Ni recoveries. These findings are important in enabling Cu–Ni concentrators to decouple and understand water quality effects in order to develop water sensitive control strategies for the desired flotation outcomes.

Keywords: Flotation performance, froth stability, gangue management, ionic strength, water quality, water recycling

3.1 Introduction

Cu–Ni bearing Nkomati ores from South Africa are exploited and concentrated by means of froth flotation. The process has a strong reliance on various physico-chemical parameters. The volume and quality of water are important variables in flotation, as water is both a transport medium, as well as a reagent. Water demand, its scarcity, and the resultant stringent legislation on water conservation and usage are becoming serious threats to the sustainability of the concentration of these ores through flotation. In order to safeguard the process from these threats, the mining industry is resorting to recycling their process water onsite or making use of other water sources, such as treated sewage water, seawater, and gray water. However, the recycling of process water results in the accumulation of various chemical and biological substances such as inorganic electrolytes and unspent residual organic reagents. These biological and chemical substances may be detrimental to the process. The presence of inorganic electrolytes in water may form either oxyhydroxo species or precipitates in the flotation pulp when the process pH is altered beyond normal operating pH of 7–9. Unfortunately, process water with high total dissolved solids (TDS) and I tends to have a high pH within the range of 10–12 [1, 2]. This change in the characteristics of the process water may alter chemical interactions that enable the process.

The need for a cleaner and safer environment with a clean surface and ground water does, however, put pressure on increasing the recycling and reuse of process water within minerals processing, despite the possible unintended consequences. Added to this is the demand for zero effluent discharge and zero freshwater withdrawal for industrial use. These challenges, even more, add to the need for mining to reuse onsite recirculated water. Since the chemistry of process water is entirely different from fresh water and changes as water is recycled, there is, however, a concern about the possible effects of its constituents, such as ion type and ionic strength, on the efficiency of the flotation process [3]. Studies have shown that recycled process water in mineral processing has high salinity and high ionic strength, and contains heavy metal ions, residual flotation reagents, organics, and other pollutants compared to fresh municipal water [4]. Therefore, process water treatment and reuse have much academic and industrial relevance where processes such as flotation are concerned [5, 6].

In response to the scarcity of fresh water, tight environmental legislation, and the need for the industry to be resilient against these challenges, mineral processing circuits are increasingly making use of onsite recycled process water. However, the chemistry of recycled process water may alter the performance of the flotation system. Therefore, it is required to understand the effects of recycled process water constituents on the flotation process. Inorganic electrolytes and their high concentrations in recycled process water are key parameters to consider since inorganic electrolytes in recycled process water can hinder the collector from adsorbing onto the mineral surface [7–12].

It has been reported that ions such as Ca^{2+}, $S_2O_3^{2-}$, and SO_4^{2-} may activate pentlandite and pyrrhotite mineral surfaces at normal process pH 9, while $S_2O_3^{2-}$ and Ca^{2+} compete with xanthate for adsorption on pentlandite surfaces [13], thus making the case for research into oxyhydroxo species in flotation, as these may affect the sub-process of flotation [11, 14, 15]. Shackleton *et al.* [16] stated that Ca^{2+} ions can chemisorb onto pentlandite surface, replacing metal ions at the surface and possibly reducing pentlandite particle surface hydrophobicity. Metal ions hydrolyze in alkaline pH and may deposit as hydrophilic metal hydroxides, sulfates, or carbonates on the mineral surfaces [17]. The reduction in mineral surface hydrophobicity could compromise the efficiency of the bubble–particle attachment sub-process of flotation, and this may consequently affect the performance of the flotation system negatively [11, 14, 18, 19]. Inorganic electrolyte type, concentration, and ionic strength in solution can have an influence on the extent of mineral activation. Thus, this may affect mineral particle hydrophobicity or hydrophilicity [20–23].

The presence of divalent cations, such as Mg^{2+} or Fe^{2+}, in the solution can compete with metal ions like Cu^{2+} on the mineral surface [24]. Previous studies have shown that the depression of naturally floatable gangue and the efficacy of carboxymethyl cellulose (CMC) as depressants were very much dependent upon the type and ionic strength of the electrolytes present in the solution [25–31]. They showed that when divalent cations, such as those of Ca^{2+} and Mg^{2+}, were present in the solution, and the depression was greater than when monovalent cations were present (e.g., K^+). It was also shown that increasing the ionic strength of the electrolytic conditions resulted in even greater depression of the naturally floatable gangue (NFG).

Certain inorganic electrolytes may improve the efficacy of CMC adsorption onto floatable gangue such as talc, thereby resulting in greater depression of NFG and an improvement of sulfide concentrate grades. The presence of inorganic electrolytes can also improve the bubble-particle attachment efficiency by compressing the electrical double layer, thus reducing the electrostatic repulsion between particles and bubbles [9, 10, 32]. Further to the aforementioned action of inorganic electrolytes, these inorganic electrolytes may also favor the inhibition of bubble coalescence, harnessing the formation of smaller and more stable bubbles due to their influence on gas dispersion properties, such as bubble size and gas hold-up [33–38]. This, in turn, would increase the stability of the froth. Smaller bubbles in flotation are also important for the fact that they increase the bubble–particle collision probability [39, 40]. This, in turn, improves bubble–particle attachment efficiency.

Fundamentally, it is known that an increase in the concentration of inorganic electrolytes increases the density of ions in the bulk solution. This results in counter-ion diffusion into the Stern layer. This increases attraction forces and decreases repulsion forces in solution [1, 41]. The impact of ionic strength on bubble size is attributed to the fact that inorganic ions in water seem to slow inter-bubble drainage, thereby inhibiting bubble coalescence and promoting the stability of the froth [36, 42, 43]. However, research on whether there is an interactive effect of the ionic strength of plant water and pH is yet to be conducted especially given that pH modifiers are inorganic electrolytes; their addition increases ion concentration, total ionic strength, and total dissolved solids. Existing literature has studied these two parameters, ionic strength and pH, individually and independently of one another until recent studies [11, 14]. This has shown the importance of studying these parameters simultaneously owing to the oxyhydroxo species that form at pH > 9 and their important effects on sub-processes of flotation, such as bubble–particle attachment. The pH of solution or pulp has been investigated mostly on single salt electrolytes in order to simplify the complex water chemistry [44]. However, this approach does not truly represent the complex water chemistry in flotation circuits and would not allow plant operators to predict the impact of ionic strength and pH when more than one oxyhydroxo species, inorganic electrolyte, ionic complex, or precipitate are present in the pulp phase. It has been shown that at pH values of the pulp above 9, the concentration of Ca^{2+}, Mg^{2+}, and Na^{+} decreases, while the concentration of their respective hydroxides, carbonates, and sulfates increases [1, 30]. The precipitation of these hydroxides, carbonates, and sulfates could produce slime coatings on the mineral surface and result in the mineral being rendered hydrophilic or the reaction of the collector on the mineral surface could possibly be hindered [40].

Therefore, the aim of the study presented in this chapter is to investigate increases in the ionic strength of synthetic plant water and increases in pulp pH, which introduces oxyhydroxo species in the solution, on the floatability of a chalcopyrite and pentlandite containing ore. This study has a keen interest on the floatability of the Cu–Ni bearing sulfide minerals, froth stability, and gangue management by studying total solids recoveries and the concentrate grades of Cu and Ni.

To complement the findings from the 3-phase batch flotation tests (consisting of solid particles–ore, liquid–water, and gas–air), 2-phase batch flotation and 2-phase froth column tests (consisting of a liquid–water and gas–air) were conducted.

3.2 Materials and Methods

All tests described hereafter were conducted in duplicates in order to ensure reproducibility and reliability of the laboratory results. Error bars representing standard error are shown on each graph presented in the Results section and the standard error was well within 5% for each data presented.

3.2.1 Three-Phase Batch Flotation

Three-phase batch flotation tests were performed at the Vaal University of Technology's Metallurgical Engineering Laboratories. Ore samples (1 kg) of a selected Cu–Ni bearing ore were milled at 66% solids in synthetic plant water for 15 min in order to obtain a grind of 60% passing 75 μm. The milled slurry was then transferred into a 3-L Denver batch flotation cell in order to perform non-sequential bulk flotation of chalcopyrite and pentlandite (Cu and Ni, respectively). The volume of the cell was made up to generate 35% solids using synthetic plant water (at the required ionic strength). The cell was fitted with a variable speed drive, and the pulp level was controlled manually. The impeller speed was set at 1200 rpm. An air flow rate of 7 L/min was maintained for all flotation experiments and a constant froth height of 2 cm was sustained throughout. Reagents were dosed and allowed to condition as shown in Table 3.1. Concentrates C1, C2, C3, and C4 were collected at 2, 6, 12, and 20 min, respectively, by scraping the froth into a collection pan every 15 sec. A feed sample was taken before each flotation test and a tails sample after each flotation test. Water usage and pH were monitored throughout the tests. Feeds, concentrates, and tails were filtered, dried, and weighed before analysis. The flotation test work was carried out at pH 9 and pH 11, with stock solutions of NaOH and HCl used for pH adjustment and control. The addition of reagents and their sequence thereof are shown in Table 3.1. Sodium isobutyl xanthate (SIBX) with a 95% active content was used as the collector and dosed into the mill at 150 g/t, allowing for its conditioning time of 15 min, as it had been previously shown by Wiese *et al.* [45] that the collector is best added and conditioned in the mill. Sendep 30 E, a polysaccharide of the carboxy methyl cellulose (CMC) type, was used as the depressant, dosed into the batch flotation cell at the dosages indicated in Table 3.1 and allowed to condition for 3 min. Collector was prepared fresh every day while depressant was prepared and used over 3 days, respectively. Dow 200, a polyglycol ether, was used as the frother, dosed into the flotation cell at 40 g/t and conditioned for 1 min before air was introduced. Synthetic plant water (SPW), shown in Table 3.2, with various ionic strengths (*I*) and total dissolved solids (TDS) was used throughout the laboratory tests. It is worth mentioning that the synthetic plant water recipe for 1 SPW was developed within the Centre for Minerals Research (CMR) to emulate or resemble the quality of process water from a typical Cu–Ni–PGM ore concentrator [45]. Owing to onsite process water recirculation, the quality of process water is said to be different from the quality of water represented by 1 SPW. To account for possible increases in inorganic ion concentrations with increased on site was recycling, SPW is multiplied 5 and 10 times, denoted as 5SPW and 10SPW [31, 46]. Copper and nickel assays were conducted at the University of Pretoria using X-Ray Fluorescence (XRF). All tests were conducted in duplicate to minimize experimental error, hence the standard error bars shown in Figures 3.1, 3.2, and 3.3.

Table 3.1 Reagent addition and sequence.

Reagent	Type	Dosages (g/t)	Conditioning time (min)
Collector	SIBX	150	15
Depressant	Sendep 30E	0, 100, 500	3
Frother	DOW200	40	1

Table 3.2 Water quality of synthetic plant water (SPW).

Water type	Ca^{2+} (mg/L)	Mg^{2+} (mg/L)	Na^{+} (mg/L)	Cl^{-} (mg/L)	SO_4^{2-} (mg/L)	NO_3^{-} (mg/L)	CO_3^{2-} (mg/L)	TDS (mg/L)	Ionic strength [mol/L]
1 SPW	80	70	153	287	240	176	17	1023	0.0242
5 SPW	400	350	765	1435	1200	880	85	5115	0.1205
10 SPW	800	700	1530	2870	2400	1760	170	10230	0.2420

3.2.2 Two-Phase Batch Flotation

For the two-phase batch flotation tests, only the water type (as shown in Table 3.2) and pH (pH 9 and pH 11) were varied. Prepared plant water solutions were transferred directly to a 3 L Barker flotation cell at the Centre for Minerals Research (CMR) at the University of Cape Town (UCT). The operating conditions of the cell were maintained as in the three-phase test work, with the frother, DOW 200, dosed at 40 ppm. The collection of (foam) concentrates C1, C2, C3, and C4 was done at 2, 6, 12, and 20 min, respectively by scrapping the foam into a collecting pan every 15 sec; this ensured that there were four (foam) concentrates collected altogether. These were weighed in order to record or account for the amount of foam reporting to the collecting pans. Figure 3.4, however, reports only the cumulative total amount of water recovered.

3.2.3 Two-Phase Froth Column

Two-phase froth column tests were conducted for all the different synthetic plant water qualities presented in Table 3.2. These tests were conducted at both pH 9 and pH 11 with a fixed frother dosage of 5 ppm DOW 200. A full experimental setup for the test rig and operation of the UCT CMR froth column, which was used to obtain the foam height and collapse time presented in this study, is reported elsewhere in Manono *et al.* [37]. All tests were conducted in duplicate to minimize experimental error, hence the standard error bars shown in Figure 3.5.

3.3 Results and Discussion

3.3.1 Solids and Water Recoveries from a Three-Phase Batch Cell

Figure 3.1 shows the total amount of solids and water recovered for all tested conditions.

From Figure 3.1, it is evident that at pH 9 and pH 11, both the solids and water recoveries decreased with an increase in depressant dosage for all ionic strength conditions. Further to this, it can be seen that with increasing ionic strength at pH 9, the recovery of solids and water was higher compared to lower ionic strengths. Interestingly, at pH 11, an opposite trend

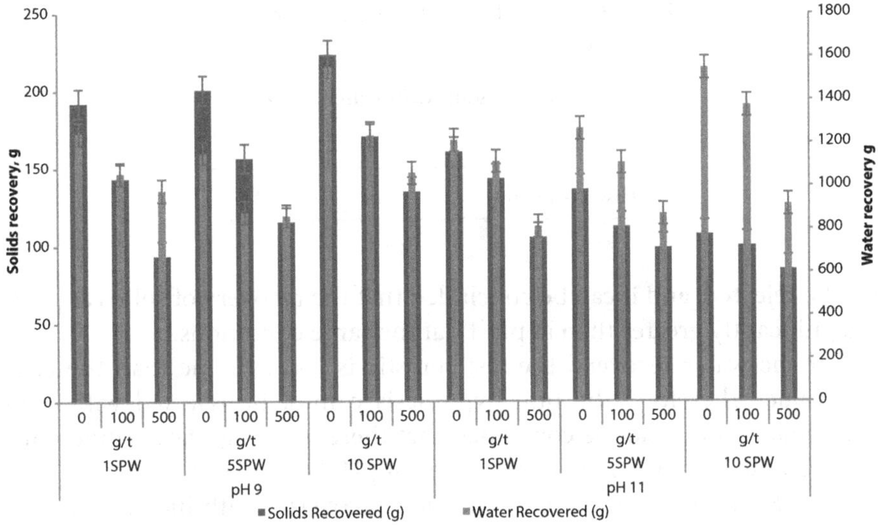

Figure 3.1 Final solids and water recoveries in increasing ionic strength of synthetic plant water, pH, and depressant dosage.

was observed regarding the effect of ionic strength on solids recovery. Here, the solids recovery decreased as the ionic strength of plant water increased. The water recovery, however, increased at increasing ionic strengths, as is the case at pH 9. However, both water and solids recoveries at pH 11 were lower than the results obtained at pH 9 for every corresponding SPW and depressant condition.

A paired t-test was performed to test whether there is a significant difference between the solids recovered at pH 9 and pH 11, and water recovered at pH 9 and pH 11. The results are given in Tables 3.3 and 3.4.

The results from the paired t-test on solids recovery show that the test statistic is more extreme than the critical value and consequently falls in the rejection region. Therefore, there is evidence with 95% confidence that the null hypothesis of no difference in the solids recoveries at pH 9 and pH 11

Table 3.3 Paired t-test for solids recoveries: pH 9 and pH 11.

Hypothesized mean difference	**0**
Test statistic	3.24
t critical one-tail	1.86

Table 3.4 Paired t-test for water recoveries: pH 9 and pH 11.

Hypothesized mean difference	**0**
Test statistic	-0.53
t critical one-tail	1.86

can be rejected, and it can be concluded that the recovery of solids at pH 9 is significantly greater than at pH 11 at the same conditions.

For the water recovery, the test statistic is less extreme than the critical value. Therefore, the null hypothesis cannot be rejected with 95% confidence, and it can be concluded that there is no significant difference between the amount of water recovered at pH 9 and pH 11.

Figure 3.1 shows an increase in water recoveries with increases in the ionic strength of plant water for both pH conditions. High water recoveries are generally translated to high froth stability [47]. The increase in water recovery with the increasing ionic strength of process water is speculated to have been caused by an enhancement in froth stability due to the increased concentration of ions in the plant water [38, 43, 48]. It is said that an increase in the concentration of ions creates a more stable froth. Furthermore, the increase in solids and water recovery (pH 9) and water recovery (pH 11), with increasing ionic strength of plant water can be attributed to the fact that inorganic ions reduce the zeta potential of the mineral and compress the electrical double layer, in turn reducing the repulsive forces between the bubble and the particle, enhancing bubble-particle attachment, and ultimately leading to higher water and solids recoveries [49–52]. The increase in water recovery at pH 11 with increasing ionic strength of plant water, despite a decrease in solids recoveries, is attributed to an increase in the ionic strength of solution in that when pH is modified from pH 9 to pH 11, additional ions are added to the system, increasing the presence of hydroxyl ions, thereby increasing the ionic strength of solution causing an increase in the mobility and stability of the froth. This is in line with the theory that increasing the ionic strength of plant water or solution and pH inhibits bubble coalescence and consequently retarding inter-bubble drainage, and stabilizing the froth [36, 53].

An interesting feature of the observations made in Figure 3.3 is that at pH 11, solids recovery decreases as the ionic strength of plant water increases, thus showing the opposite effect on solids recovery when compared to pH 9, where solids recovery increased with increasing ionic strength of plant water. The decrease in solids recovery at pH 11 with increasing ionic strength is believed to have resulted from the presence of ions, oxyhydroxo species, and precipitates that influence the mineral surface by changing the hydrophobicity of the mineral and hindering the collector from adsorbing onto the mineral surfaces [40]. At alkaline conditions, collector ions compete with OH^- and oxyhydroxo species formed in solution to adsorb on the surface of the mineral [11, 27]. In addition, speciation diagrams of these synthetic plant waters reported in Manono *et al.* [1, 30] provide evidence that as the pH of the solution increases, the concentration of ions such as Na^+, Ca^{2+}, and Mg^{2+} decreases, and hydroxide, carbonate, and sulfate precipitates are formed. It is reported that these substances tend to form slime coatings on the surface of the mineral, changing the mineral surface chemistry, thus hindering the adsorption of the collector on the mineral surface, and thereby preventing flotation [11, 54]. It can therefore be deduced that a higher concentration of dissolved ions and increased pH in the flotation system facilitated a greater change in the surface chemistry of the mineral, thus preventing the action of the collector. This result is also consistent with the findings of Tadie *et al.* [55], which revealed that the recovery of galena decreased significantly at pH 11.8 compared to pH 9.2 conditions, reinforcing the depressing effect of oxyhydroxo species at high pH conditions. Additionally, Ikumapayi *et al.* [56] showed that in the presence of calcium and sulfate ions, the recovery of galena decreased due to the presence of inorganic ions changing the zeta potential of the mineral and consequently hindering the adsorption of xanthate on the surface of the mineral.

3.2 Cu and Ni Recoveries and Grades from a Three-Phase Batch Cell

Figure 3.2 illustrates the final Cu recovery and grade respectively for all the conditions tested in this study. For each SPW type and at both pH 9 and 11, the recovery of Cu decreased with an increase under the depressant dosage, while Cu grades showed an increase with increasing depressant dosages. Though chalcopyrite is a fast-floating mineral, it seems that the effectiveness of CMC as a gangue depressant not only reduced the recovery of gangue, but it also reduced that of chalcopyrite though not to a great extent. This is expected in complex mineral systems where there may be unliberated or composite particles

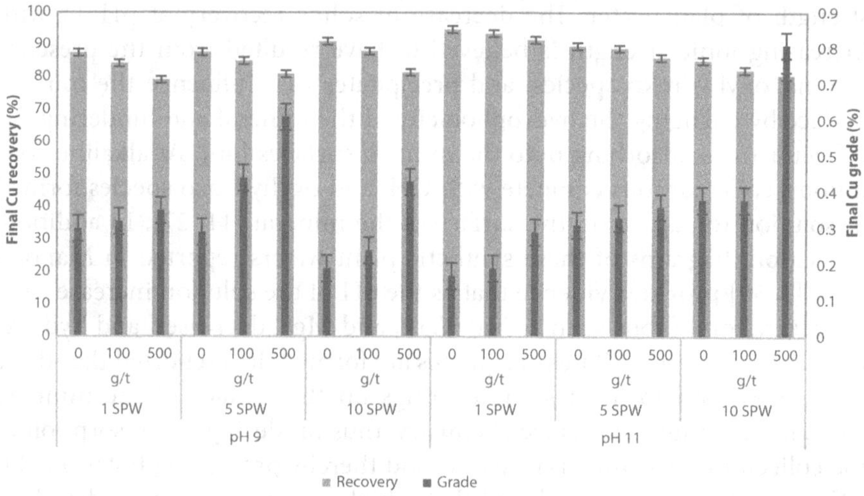

Figure 3.2 Final Cu recovery and grade in increasing ionic strength of synthetic plant water, pH, and depressant dosage.

present [57–59]. At pH 9 and at a fixed depressant dosage, increasing the ionic strength of SPW did not affect Cu recoveries. However, Cu grades decreased between 1 SPW and 10 SPW, suggesting that increases in the ionic strength had no effect on the behavior of chalcopyrite but had an increasing effect on the recoveries of gangue minerals owing to the increasing froth stability seen from the results in Figure 3.1. However, 5 SPW resulted in grades higher than those obtained in 1 SPW and 10 SPW. This behavior with 5 SPW has been seen before on pyrrhotite and a Cu–Ni–PGM Merensky ore, and speculations are that the solution chemistry at 5 SPW could be responsible for this trend. This, however, needs further investigative work [9, 31].

Figure 3.3 depicts the final Ni recovery and grade. At pH 9, it is shown that an increase in depressant dosage resulted in a decrease in Ni recovery, as expected. An increase in the ionic strength of synthetic plant water also resulted in a decrease in Ni recovery as shown in Figure 3.1. Ni grades increased with an increase in depressant dosage, as expected. However, there was a decrease in mineral grades with increasing ionic strength of synthetic plant water. Ni recoveries are lower when the pH is increased from pH 9 to pH 11. It is also worth noting that at pH 11, mineral (pentlandite) grades tend to increase with increasing ionic strength of plant water, contrary to the trend seen at pH 9.

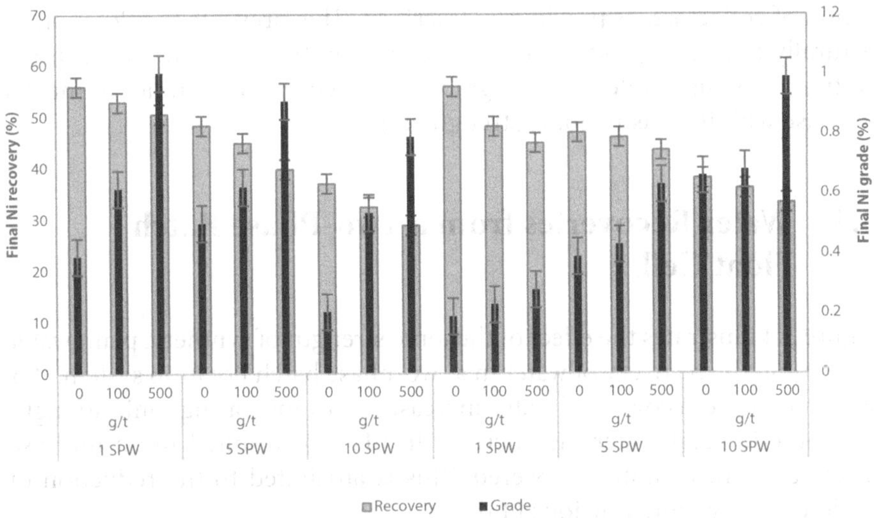

Figure 3.3 Final Ni recovery and grade in increasing ionic strength of synthetic plant water, pH and depressant dosage.

The key findings from Figures 3.2 and 3.3 at pH 9 are that the copper and nickel grades decreased as the ionic strength increased lower coalescence, smaller bubbles, larger surface area for attachment and entrainment. This phenomenon can be explained by the froth stabilizing effect of the ions present in solution. A higher ionic strength, and therefore greater froth stability, allow for more of the gangue material to be recovered to the concentrate through entrainment [31]. The opposite trend was, however, observed at pH 11. Here, both the copper and nickel grades increased with increasing ionic strength. This result could be because of the fact that at high ionic strengths, the concentrations of Ca^{2+} and Mg^{2+} are higher compared to those at low levels of ionic strength. This was clearly shown by speciation of selected synthetic plant waters [1]. As shown by Shortridge *et al.* [26], in the presence of these ions, the depressing ability of CMC is enhanced because of the strong negative charge density of this depressant. Moreover, Parolis *et al.* [25] showed that divalent metal cations, such as $Ca^{2+}$and Mg^{2+}, assisted in the adsorption of CMCs on to gangue such as talc because a greater coiling of the CMC chain on the surface of the adsorbed mineral (talc) is achieved, thus, improving the efficacy of CMC adsorbing onto talc. It can thus be deduced that in high concentrations of these ions, more naturally floating gangue could have been depressed. Furthermore, the high level of alkalinity in the system results in the formation of hydroxide complexes that coat mineral surfaces and hinder the action of collectors, as previously described. This prevents the flotation of naturally floatable gangue such as talc. Taking these factors into consideration, it is justifiable that the grade and recoveries of sulfide minerals increase with increasing ionic strength at pH 11.

3.3 Water Recoveries from a Two-Phase Batch Float Cell

Figure 3.4 illustrates the effect of the ionic strength of synthetic plant water and pH on the recovery of water in a two-phase batch flotation system. It is shown that the recovery of water increased with increasing ionic strength of synthetic plant water. An increase in pH also resulted in an increase in the amount of water recovered. This is attributed to the reduction of coalescence by inorganic ions [43].

3.4 Froth Column Studies from a Two-Froth Column

Figure 3.5 illustrates the effects of the ionic strength of synthetic plant water and increasing pH on foam height and foam collapse time as these measures are typically used when assessing the stability of the froth [60–63]. Both foam height and foam collapse time increased with increasing ionic

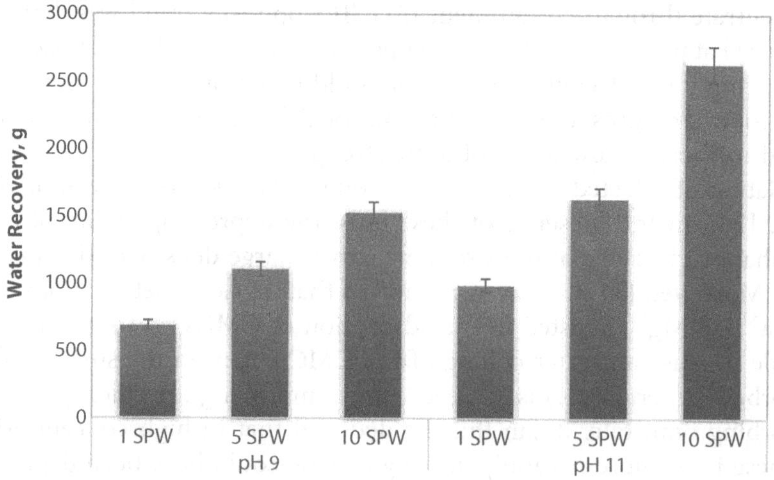

Figure 3.4 The effects of ionic strength of synthetic plant water and pH on the recovery of water.

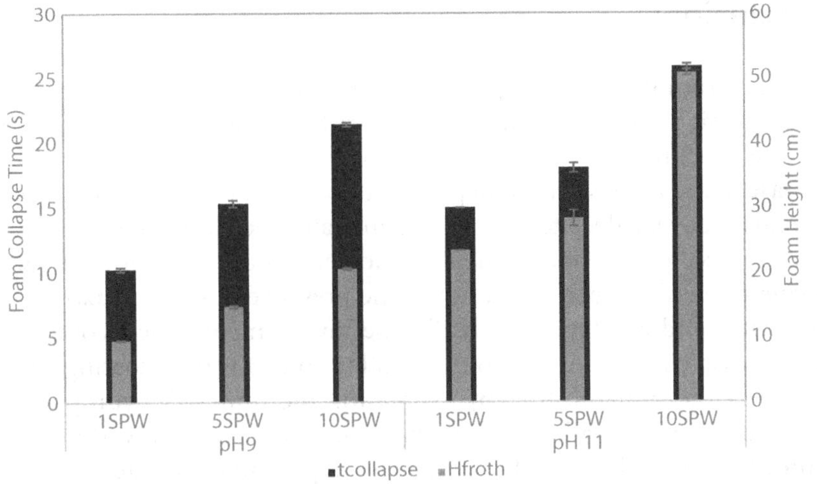

Figure 3.5 The effect of the ionic strength of synthetic plant water and pH on foam height and foam collapse time.

strength and pH. The higher pH resulted in higher foam height and longer foam collapse time compared to the lower pH.

Experimental evidence of an increase in both solids and water recoveries at pH 9, and the increases in water recoveries at pH 11 with increasing ionic strength of plant water was attributed to an increase in froth stability. As a complement to this claim of increased froth stability, a frother dosage of 5 ppm was chosen in the two-phase experiments so as to elucidate frothing effects of the ionic strength of synthetic plant water and pH, as this dosage is below the selected frother's critical coalescence concentration. Thus, enabling any effects resulting from increasing ionic strength of SPW and oxyhydroxo species to be revealed. The results shown in Figure 3.5 are indicative of an enhancement in froth stability with increasing ionic strength of SPW and pH and are in agreement with both the two-phase and three-phase batch flotation trends on water recoveries shown in Figure 3.1 and Figure 3.4, respectively. The increase in bubble coalescence retarding inorganic electrolytes such as those associated with calcium, magnesium, and sulfate ions results in the decrease in bubble coalescence and thus decrease in bubble size. It is thus postulated that this increase in bubble coalescence retarding inorganic electrolytes results in a froth stabilizing effect with pH and ionic strength increases. The mechanism for this phenomenon is attributed to the increase in concentration of these inorganic electrolytes, having an ability to decrease inter-bubble drainage rate and the stability of the inter-bubble film [36].

Conclusions

It is clear from the results reported in this study that the increase in both pH and ionic strength of plant water favors higher water recoveries, which indicate an improvement in froth stabilization. It was also shown that an increase in the ionic strength of plant water at pH 11 resulted in a decrease in solids recovery despite an increase in water recovery. This finding can be attributed to the enhancement of the depression of naturally floatable gangue minerals, which is a result of the presence of oxyhydroxo species, carbonates, and sulfates in highly alkaline environments. These compounds are depressive in nature due to their ability to form slime coatings on the surface of the mineral, thereby decreasing the mineral's hydrophobicity and floatability. Overall, this study has shown that the inadvertent effects of process water salinity and irregularities in pH can be countered by using higher depressant dosages, and still achieve good flotation performance as regards valuable mineral grades and recoveries. This could have serious implications on process water treatment cost savings in that a removal or reduction of suspect ions such as SO_4^{2-} and Ca^{2+} might not be necessary, but the reagent chemistry could be channelled in order to address their presence in the flotation process water and their possible effects on flotation performance. It is important to note that the effects seen in this work could very well be ore-specific, thus future research should consider process water quality effects on different ore types.

Acknowledgments

The authors would like to thank AECI Senmin, the University of Cape Town, and the Vaal University of Technology for their respective contributions towards this work. This work is supported in part by the National Research Foundation (NRF) of South Africa (Grant numbers 99262 and 103641). Any opinions, findings, conclusions, or recommendations expressed in any publication generated by NRF-supported research are those of the authors and the NRF accepts no liability whatsoever in this regard. Financial and technical contributions from the South African Minerals to Metals Research Institute (SAMMRI) and its members are also acknowledged. The authors would also like to thank the reviewers who provided feedback and comments that helped in improving the original manuscript of this work.

References

1. Manono, M., Corin, K., Wiese, J., Perspectives from literature on the influence of inorganic electrolytes present in plant water on flotation performance. 54, 1191–1214, 2018.
2. Michaux, B., Rudolph, M., Reuter, M.A., challenges in predicting the role of water chemistry in flotation through simulation with an emphasis on the influence of electrolytes. *Miner. Eng.*, 125, 252–264, 2018.
3. Rao, S.R. and Finch, J.A., A review of water re-use in flotation. *Miner. Eng.*, 2, 65–85, 1989.
4. Slatter, K.A., Plint, N.D., Cole, M., Dilsook, V., De Vaux, D., Palm, N., Oostendorp, B., Water management in anglo platinum process operations : Effects of water quality on process operations. *International Mine Water Conference*, Pretoria, South Africa, October 19th–23rd, 2009 Proceedings, pp. 46–55, ISBN: ISBN Number: 978-0-9802623-5-3, 2009.
5. Chen, J., Liu, R., Sun, W., Qiu, G., Effect of mineral processing wastewater on flotation of sulfide minerals. *Trans. Nonferrous Met. Soc. China*, 19, 454–457, 2009.
6. Michaux, B., Hannula, J., Rudolph, M., Reuter, M.A., van den Boogaart, K.G., Möckel, R., Kobylin, P., Hultgren, M., Peltomäki, M., Roine, A. *et al.*, Water-saving strategies in the mining industry–the potential of mineral processing simulators as a tool for their implementation. *J. Environ. Manage.*, 234, 546–553, 2019.
7. Hancer, M., Celik, M.S., Miller, J.D., The significance of interfacial water structure in soluble salt flotation systems. *J. Colloid Interface Sci.*, 235, 150–161, 2001.
8. Wang, B. and Peng, Y., The effect of saline water on mineral flotation–A critical review. *Miner. Eng.*, 13–24, 2014.
9. October, L., Corin, K., Schreithofer, N., Manono, M., Wiese, J., Water quality effects on bubble-particle attachment of pyrrhotite. *Miner. Eng.*, 131, 2019.
10. October, L.L., Corin, K.C., Manono, M.S., Schreithofer, N., Wiese, J.G., A fundamental study considering specific ion effects on the attachment of sulfide minerals to air bubbles. *Miner. Eng.*, 2020.
11. October, L.L., Manono, M.S., Corin, K.C., Schreithofer, N., Wiese, J.G., The influence of specific ions and oxyhydroxo species in plant water on the bubble-particle attachment of pyrrhotite. *ACS Omega*, 6, 28496–28506, 2021.
12. Manenzhe, R.M., Corin, K.C., Wiese, J.G., Manono, M.S., Musuku, B., The effect of water quality on the adsorption of a xanthate collector in the flotation of a sulphide ore, in: *Proceedings of the IMPC 2018-29th International Mineral Processing Congress*, 2019.
13. Hodgson, M. and Agar, G.E., Eilectrochemical investigations into the flotation chemistry of pentlandite and pyrrhotite: Process water and xanthate interactions. *Can. Metall. Q.*, 28, 189–198, 1989.

14. October, L.L., Corin, K.C., Manono, M.S., Schreithofer, N., Wiese, J.G., *Considering The Ionic Strength And Ph of Process Water on Bubble-Particle Attachment of Sulfide Minerals : Implications For Froth Flotation in Saline Water*, pp. 437–444, 2001.
15. Muzenda, E., Afolabi, A.S., Abdulkareem, A.S., Ntuli, F., Effect of PH on the recovery and grade of base metal sulphides (PGMs) by flotation. *Proceedings of the World Congress on Engineering and Computer Science 2011*, vol. II, pp. 19–22, 2011.
16. Shackleton, N.J., Malysiak, V., O'Connor, C.T., The use of amine complexes in managing inadvertent activation of pyroxene in a pentlandite–pyroxene flotation system. *Miner. Eng.*, 16, 849–856, 2003.
17. Fuerstenau, D.W., Herrera-Urbina, R., McGlashan, D.W., Studies on the applicability of chelating agents as universal collectors for copper minerals. *Int. J. Miner. Process.*, 2000.
18. Koh, P.T.L., Hao, F.P., Smith, L.K., Chau, T.T., Bruckard, W.J., The effect of particle shape and hydrophobicity in flotation. *Int. J. Miner. Process.*, 93, 128–134, 2009.
19. Schwarz, S. and Grano, S., Effect of particle hydrophobicity on particle and water transport across a flotation froth. *Colloids Surf. A Physicochem. Eng. Asp.*, 256, 157–164, 2005.
20. Dzingai, M., Manono, M.S., Corin, K.C., Probing the effect of water recycling on flotation through anion spiking using a low-grade cu–ni–pgm ore: The effect of NO3–, SO42– and S2O32–. *Minerals*, 11, 340, 2021.
21. Dzingai, M., Manono, M., Corin, K., Simulating the effect of water recirculation on flotation through ion-spiking: Effect of Ca2+ and Mg2+. *Minerals*, 10, 1033, 2020.
22. Manono, M.S., Corin, K.C., Wiese, J.G., The influence of electrolytes present in process water on the flotation behaviour of a Cu-Ni containing ore. *Miner. Eng.*, 96–97, 2016.
23. Manono, M.S., Corin, K.C., Wiese, J.G., The behavior of gangue during the flotation of a sulfidic pgm-bearing ore in response to various monovalent and divalent ions in process water. *Front. Chem.*, 2020.
24. Lascelles, D., Sui, C.C., Finch, J.A., Butler, I.S., Copper ion mobility in sphalerite activation. *Colloids Surf. A Physicochem. Eng. Asp.*, 186, 163–172, 2001.
25. Parolis, L.A.S., van der Merwe, R., Groenmeyer, G.V., Harris, P.J., The influence of metal cations on the behaviour of carboxymethyl celluloses as talc depressants. *Colloids Surf. A Physicochem. Eng. Asp.*, 317, 109–115, 2008.
26. Shortridge, P.G., Harris, P.J., Bradshaw, D.J., The influence of ions on the effectiveness of polysaccharide depressants in the flotation of Talc, in: *Proceedings of the Polymers in Mineral Processing. 3rd UBC-McGill Bi-Annual International Symposium on Fundamentals of Mineral Processing*, CIM, Quebec City, Canada, pp. 155–169, 1999.

27. Burdukova, E., Van Leerdam, G.C., Prins, F.E., Smeink, R.G., Bradshaw, D.J., Laskowski, J.S., Effect of calcium ions on the adsorption of CMC onto the basal planes of New York Talc-a ToF-SIMS study. *Miner. Eng.*, 21, 1020–1025, 2008.
28. Corin, K.C., Reddy, A., Miyen, L., Wiese, J.G., Harris, P.J., The effect of ionic strength of plant water on valuable mineral and gangue recovery in a platinum bearing ore from the merensky reef. *Miner. Eng.*, 24, 131–137, 2011.
29. Manono, M.S., Corin, K.C., Wiese, J.G., Inorganic electrolytes on the efficacy of a carboxymethyl cellulose as a coagulant for talc : Implications for talc depression in flotation, in: *Proceedings of the Mine Water–Technological and Ecological Challenges*, International Mine Water Association, Perm, Russia (Perm State University, pp. 217–224, 2019.
30. Manono, M., Corin, K., Wiese, J., The effect of the ionic strength of process water on the interaction of talc and cmc: Implications of recirculated water on floatable gangue depression. *Minerals*, 9, 2019.
31. Manono, M., Corin, K., Wiese, J., Water quality effects on a sulfidic PGM ore : Implications for froth stability and gangue management. 54, 1253–1265, 2018.
32. Kurniawan, A.U., Ozdemir, O., Nguyen, A.V., Ofori, P., Firth, B., Flotation of coal particles in MgCl2, NaCl, and NaClO 3 solutions in the absence and presence of dowfroth 250. *Int. J. Miner. Process.*, 2011.
33. Quinn, J.J., Kracht, W., Gomez, C.O., Gagnon, C., Finch, J.A., Comparing the effect of salts and frother (MIBC) on gas dispersion and froth properties. *Miner. Eng.*, 20, 1296–1302, 2007.
34. Quinn, J.J., Sovechles, J.M., Finch, J.A., Waters, K.E., Critical coalescence concentration of inorganic salt solutions. *Miner. Eng.*, 58, 1–6, 2014.
35. Craig, V.S.J., Bubble coalescence and specific-ion effects. *Proceedings of the Current Opinion in Colloid and Interface Science*, 2004.
36. Craig, V.S.J., Ninham, B.W., Pashley, R.M., The effect of electrolytes on bubble coalescence in water. *J. Phys. Chem.*, 97, 10192–10197, 1993.
37. Manono, M.S., Corin, K.C., Wiese, J.G., The effect of ionic strength of plant water on foam stability: A 2-phase flotation study. *Miner. Eng.*, 40, 42–47, 2013.
38. Manono, M.S. and Corin, K.C., Considering specific ion effects on froth stability in sulfidic Cu-Ni-PGM ore flotation. *Minerals*, 12, 2022.
39. Pugh, R.J., Weissenborn, P., Paulson, O., Flotation in inorganic electrolytes, the relationship between recover of hydrophobic particles, surface tension, bubble coalescence and gas solubility. *Int. J. Miner. Process.*, 51, 125–138, 1997.
40. Biçak, Ö., Ekmekçi, Z., Can, M., Öztürk, Y., The effect of water chemistry on froth stability and surface chemistry of the flotation of a Cu-Zn sulfide ore. *Int. J. Miner. Process.*, 102–103, 32–37, 2012.
41. Wang, B. and Peng, Y., The effect of saline water on mineral flotation-a critical review. *Miner. Eng.*, 66, 13–24, 2014.

42. Barker, L.M., *The Effect of Electrolytes on the Flotation of Pyrite*, University of Cape Town, 1986.
43. Manono, M.S., Corin, K.C., Wiese, J.G., The effect of ionic strength of plant water on foam stability: A 2-phase flotation study. *Miner. Eng.*, 40, 2013.
44. Muzenda, E., Afolabi, A.S., Abdulkareem, A.S., Ntuli, F., Effect of PH on the recovery and grade of base metal sulphides (PGMs) by flotation. *Proceedings of the World Congress on Engineering and Computer Science 2011*, vol. II, pp. 19–22, 2011.
45. Wiese, J., Harris, P., Bradshaw, D., The influence of the reagent suite on the flotation of ores from the merensky reef. *Miner. Eng.*, 18, 189–198, 2005.
46. Corin, K.C. and Wiese, J.G., Investigating froth stability: A comparative study of ionic strength and frother dosage. *Miner. Eng.*, 66, 130–134, 2014.
47. Wiese, J.G., Harris, P.J., Bradshaw, D.J., The effect of increased frother dosage on froth stability at high depressant dosages. *Miner. Eng.*, 23, 1010–1017, 2010.
48. Corin, K.C. and Wiese, J.G., Investigating froth stability: A comparative study of ionic strength and frother dosage. *Miner. Eng.*, 66, 130–134, 2014.
49. Ejtemaei, M., Plackowski, C., Nguyen, A.V., The effect of calcium, magnesium, and sulphate ions on the surface properties of copper activated sphalerite. *Miner. Eng.*, 89, 42–51, 2016.
50. Yoon, R.H. and Sabey, J.B., Coal flotation in inorganic salt solution, in: *Interfacial Phenomena in Coal Technology*, G.D. Botsaris, and J.M. Glazmin, (Eds.), pp. 87–114, Marcel Dekker, New York, NY, USA, 1989.
51. Pugh, R.J., Weissenborn, P., Paulson, O., Flotation in inorganic electrolytes, the relationship between recover of hydrophobic particles, surface tension, bubble coalescence and gas solubility. *Int. J. Miner. Process.*, 51, 125–138, 1997.
52. Paulson, O. and Pugh, R.J., Flotation of inherently hydrophobic particles in aquoeus solutions of inorganic electrolytes. *Langmuir*, 12, 4808–4813, 1996.
53. Yousef, A.A., Arafa, M.A., Ibrahim, S.S., Abdel Khalek, M.A., Seawater usage in flotation for minerals beneficiation in arid regions-simulation and application, in: *Proceedings of the Proceedings of the 22nd International Mineral Processing Congress*, vol. 2 South African Institute of Mining and Metallurgy, Cape Town, pp. 1023–1031, 2003.
54. Rao, F., Lázaro, I., Ibarra, L.A., Solution chemistry of sulphide mineral flotation in recycled water and sea water: A review. *Miner. Process. Extr. Metall.*, 1–7, 2016.
55. Tadie, M., Corin, K., Wiese, J., O'Connor, C., Application of potential control to the flotation of galena by varying dissolved oxygen, PH and ionic strength levels, in: *Proceedings of the IMPC 2016-28th International Mineral Processing Congress*, 2016, September 2016.
56. Ikumapayi, F., Makitalo, M., Johansson, B., Rao, K.H., Recycling of process water in sulphide flotation: Effect of calcium and sulphate ions on flotation of galena. *Miner. Eng.*, 39, 77–88, 2012.

57. Corin, K.C., Reddy, A., Miyen, L., Wiese, J.G., Harris, P.J., The effect of ionic strength of plant water on valuable mineral and gangue recovery in a platinum bearing ore from the merensky reef. *Miner. Eng.*, 24, 131–137, 2011.
58. Becker, M., Harris, P.J., Wiese, J.G., Bradshaw, D.J., Mineralogical characterisation of naturally floatable gangue in merensky reef ore flotation. *Int. J. Miner. Process.*, 246–255, 2009.
59. Becker, M., De Villiers, J., Bradshaw, D., The flotation of magnetic and non-magnetic pyrrhotite from selected nickel ore deposits. *Miner. Eng.*, 2010.
60. Barbian, N., Hadler, K., Ventura-Medina, E., Cilliers, J.J., The Froth Stability Column: Linking Froth Stability and Flotation Performance. *Miner. Eng.*, *18*, 317–321, 2005.
61. Barbian, N., Ventura-Medina, E., Cilliers, J.J., Dynamic Froth Stability in Froth Flotation. *Miner. Eng.*, *16*, 1111–1116, 2003.
62. Barbian, N., Hadler, K., Cilliers, J.J., The froth stability column: Measuring froth stability at an industrial scale. *Miner. Eng.*, 19, 713–718, 2006.
63. Bikerman, J., *Foams. Applied Physics and Engineering*, vol. 10, 1973.

4

Relating the Flotation Response of Pyrrhotite to the Adsorption of Sodium Carboxymethyl Cellulose and Sodium Isobutyl Xanthate on its Surface in Process Water of a Degrading Quality

Malibongwe S. Manono*, Lisa L. October and Kirsten C. Corin

Centre for Minerals Research, Department of Chemical Engineering, University of Cape Town, Rondebosch, South Africa

Abstract

The optimization and process developments surrounding the beneficiation of Cu–Ni–PGM ores are of fundamental importance when considering the continued rise in global demand for these precious metals. The selective sequential flotation of these Cu–Ni–PGM ores is made complex by various processing challenges, such as the successful separation of the Cu–Ni–PGMs from non-sulfide naturally floatable gangue and pyrrhotite. Reagents such as sodium isobutyl xanthate (SIBX) and sodium carboxymethyl cellulose (CMC) have proven to significantly improve the selectivity of the flotation process in their respective roles. The use of recycled water with abnormal inorganic electrolyte concentrations can modify the selectivity of these reagents, as ions may adsorb onto minerals such as pyrrhotite, thereby hindering or promoting the hydrophobicity and floatability of pyrrhotite. Process water quality effects on the behavior of pyrrhotite may have a negative or positive influence on the recovery and grades of the wanted Cu–Ni–PGMs because valuable minerals associated with pyrrhotite, such as pentlandite, may be affected in like manner. Thus, the impact of concentrator process water quality on the behavior of pyrrhotite needs to be determined. In this study, synthetic process water at various ionic strengths was investigated against various CMC dosages to understand the behavior of pyrrhotite during the flotation of a Cu–Ni–PGM

**Corresponding author*: malibongwe.manono@uct.ac.za

Elvis Fosso-Kankeu, Bhekie B. Mamba and Antoine F. Mulaba-Bafubiandi (eds.) Recovery of Values from Low-Grade and Complex Minerals: Development of Sustainable Processes, (87–110) © 2024 Scrivener Publishing LLC

bearing Merensky ore in a batch flotation cell. The effect of synthetic plant water (SPW) of increasing ionic strength on SIBX adsorption on pyrrhotite was also investigated through adsorption and attachment time studies. Microflotation tests on pyrrhotite were also performed in the absence and presence of CMC in SPW of increasing ionic strength to further understand the effects on the floatability of pyrrhotite. In order to understand the effect on the mineral surface's electrokinetic potential before the addition of reagents, zeta potential determination tests were conducted.

***Keywords*:** Depressant, froth flotation, ionic strength, pyrrhotite, water quality, xanthate

4.1 Introduction

Platinum group metals (PGMs) are precious metals, which are used in catalysis, automotive, medical, and industrial manufacturing processes due to their purity, high melting points, unique catalytic properties, and high corrosion resistance [1]. The availability of these metals has become especially important and are crucial role players in power generation processes such as fuel cells or hydrogen energy that are meant to respond to the transition to a green economy. South Africa has rich natural resource endowments, and the mining sector contributes significantly to the country's economy. South Africa has the world's largest reserves of PGMs (91% of global reserves). The mining of these PGMs is a crucial part of the South African economy, as the country boasts the world's largest PGM deposit, located in the Bushveld Igneous Complex (BIC). The mining industry accounts for about 17% of South Africa's gross domestic product (GDP). These PGM ores are associated with base metal sulfides (e.g., chalcopyrite, pentlandite, and pyrrhotite).

Given that 40%–45% of the base metal sulfides (BMS) is pyrrhotite and that the surface chemistry of pyrrhotite is very sensitive to the chemistry of flotation, pyrrhotite, therefore, becomes an important sulfide to investigate in terms of its behavior. Moreover, the BIC is explored for other valuable metals such as nickel and copper, as these metals occur in association with the PGMs as BMS, namely, chalcopyrite ($CuFeS_2$) and pentlandite ($Fe_9Ni_9S_8$). However, the presence of pyrrhotite ($Fe_{(1-x)}S$) within these ore deposits provides a challenge for mineral concentrators during liberation and separation processes such as comminution and flotation [2–4].

The management of sulfide gangue for these PGM ores during flotation is thus of paramount importance as the recovery of gangue to the concentrate presents a challenge on the grade of the concentrate. Poor concentrate grades during the flotation of these PGM ores are usually attributed to the recovery of naturally floatable non-sulfide gangue, such as talc, as well as non-value bearing sulfide minerals, such as pyrrhotite, as these tend to have an affinity or association with pentlandite and other value bearing minerals, and may thus report to the concentrate through forming composites with the minerals that have been hydrophobized [3]. Managing the behavior of pyrrhotite is thus of critical importance, especially for concentrators that are grade-driven. Consequently, there is a need to understand the behavior of non-value bearing minerals such as pyrrhotite during the flotation of Cu–Ni–PGM ores. It is, however, well known that flotation as a separation technique by which these Cu–Ni–PGM ores are concentrated is reliant on a number of parameters or factors. Thus, managing any aspect of this process may not be too trivial a task, and careful consideration must be given to as many critical process parameters as possible so that the changing parameters are more manageable and enable the achievement of the desired recoveries and grades.

Water is one of these parameters, given its critical role in flotation, as it acts as both a reagent as well as a transport medium. The sensitivity of flotation to reagent chemistry and water quality necessitates that BMS contained in these Cu–Ni–PGM ores be investigated against water quality variations as the behavior of these BMS builds insight into how these valuable minerals locked up in these Cu–Ni–PGM ores would behave.

But how do these variations in water quality in Cu–Ni–PGM concentrators come about? Water variations in the industry are caused by several factors, such as the use of gray water, onsite water recirculation, treated sewage water, and seasonal water quality changes. Water scarcity, environmental concerns around the usage of freshwater resources for industrial purposes, legal implications, among others, have become major reasons behind the use of alternative water resources other than fresh municipal water. As a result of these challenges, mining operations are recycling and reusing their water to keep to a reasonably known water quality. This approach may, however, bring deleterious substances as concentrations of ions accumulate in recycled water [5, 6]. Therefore, this begs the question: What would the effect of these be on the behavior of minerals such as pyrrhotite? Because water is both a reagent and transport medium in flotation, it has been shown that changes in its quality can significantly affect flotation performance through electrolyte–reagent–mineral interactions [7, 8].

On the behavior of pyrrhotite, it has been reported that a great amount of heavy metal ions such as Cu^{2+}, Zn^{2+}, and Pb^{2+} adsorb onto the highly anionic pyrrhotite [2]. This adsorption is said to occur via an attractive force between the divalent cations and the negatively charged surface of pyrrhotite. The negatively charged collector ligand then adsorbs onto the passivated pyrrhotite, thereby inducing its hydrophobicity. However, Hodgson and Agar [9] reported that Ca^{2+}, $S_2O_3^{2-}$, and SO_4^{2-} were found to be active on the pyrrhotite surface at the normal process pH. It was found that Ca^{2+} competed with xanthate for adsorption on the surface of pyrrhotite, rendering pyrrhotite less hydrophobic, such that xanthate dosage had to be increased to impart the necessary hydrophobicity.

October *et al.* [10] investigated the bubble—particle attachment process of pyrrhotite in synthetic plant water (SPW) with an increasing ionic strength. It was shown that the attachment probability of pyrrhotite decreased as the ionic strength of synthetic plant water increased. This was further supported by a decrease in xanthate adsorption on the pyrrhotite surface with increasing ionic strength of synthetic plant water. October *et al.* [11] further investigated oxyhydroxo species and specific ions in synthetic plant water on the bubble-particle attachment of pyrrhotite. They showed that the presence of oxyhydroxo species and divalent ions, in particular, decreased the attachment of pyrrhotite particles onto bubbles. These findings were further corroborated by more fundamental tests that showed that the adsorption of the collector onto pyrrhotite decreased in the presence of oxyhydroxo species and divalent cations, suggesting that these reduced the hydrophobicity of pyrrhotite. Furthermore, microflotation tests under the same water quality conditions showed decreases in pyrrhotite recoveries in the presence of divalent ions and in increasing ionic strength of SPW.

This begs the question as to whether any competition exists between the anionic collector and the anionic depressant, as regards their adsorption onto pyrrhotite with active inorganic electrolytes from the process water, and how the accumulation of inorganic electrolytes in process water through water recycling may affect the interactions with these reagents.

In the flotation of sulfidic ores, such as the Merensky type, the targeted base metal sulfides (BMS) (chalcopyrite, pentlandite, pyrrhotite, etc.) make up about 1% of the ore, the rest being gangue (floatable and non-floatable), and 40%–45% of this BMS content is mostly in the form of pyrrhotite. There exist pyrrhotite–pentlandite and pyrrhotite–PGM composites that make the floatability of pyrrhotite a topic of interest in Cu–Ni–PGM concentrators [3, 12–17]. Given water quality variation in Cu–Ni–PGM concentrators, it is, thus, necessary to examine the behavior of pyrrhotite under varying ionic strengths of plant water.

This study involved conducting a series of batch flotation tests under various ionic strengths of process water and depressant dosages, using sodium carboxy methyl cellulose (CMC) to assess the behavior of pyrrhotite under degrading water quality during the flotation of a Cu–Ni–PGM ore from the Merensky reef. A careful mass balance was carried out to determine pyrrhotite recoveries and grades. This was made possible by analyzing the % grade and recovery of elemental copper, nickel, and sulfur, and thus, deducing the recovery of pyrrhotite, and the amount of non-sulfide gangue recovered by entrainment and true flotation. In order to investigate the flotation response of pyrrhotite from a selected PGM ore, various laboratory tests were performed. Batch flotation tests of a selected PGM ore were conducted varying the ionic strength of plant water. More fundamental tests aimed at corroborating the batch flotation tests were conducted; these included, among others, microflotation tests, bubble–particle attachment measurements using an apparatus housed at Aalto University [18, 19], collector and depressant adsorption studies, and zeta potential measurements.

4.2 Experimental Methods

4.2.1 Synthetic Plant Water Preparation

Standard synthetic plant water (1 SPW) was used as the base water quality. The levels of each ion in 1 SPW were increased 3, 5, and 10 times for the higher ionic strength flotation tests as shown in Table 4.1. This was done in order to mimic the quality of recycled water. All the chemical salts used to make up the synthetic plant water solutions were supplied by Merck and were of analytical grade.

The composition of 1 SPW shown in Table 4.1 is based on an actual PGM concentrator and was developed in the early 2000's by Wiese *et al.* [4].

Owing to the recirculation of water and in order to investigate the effect of recirculation of process water, ion concentrations were multiplied by 3, 5, and/or 10 times for worst-case scenarios. This assumed that all ion concentrations increased by the same magnitude; however, this may not always be the case for all PGM sites. However, based on plant reports and data on water quality, 3 SPW seems to be within current onsite water chemical spectra.

4.2.2 Collector Preparation

The thiol collector used in this study was sodium isobutyl xanthate (SIBX) of 97% purity supplied by AECI (Senmin). Distilled water was used to prepare a 1% (w/v) SIBX solution—fresh collector solutions were prepared daily due to the decomposing nature of xanthates.

4.2.3 Depressant Preparation

The polysaccharide used was sodium carboxymethyl cellulose, Depramin 267 (supplied by AKZO Nobel Functional Chemicals), with a molecular weight of 325 000 g/mol, a density of 1.43 g/mL, a degree of substitution of 0.62, and purity of 72%. Distilled water was used to prepare 1% (w/v) solutions of CMC. In order to ensure complete hydration of the CMC in solution, the 1% CMC solution was stirred for 2 h.

4.2.4 Ore Preparation and Milling

A Merensky reef Cu–Ni–PGM ore obtained from one of the PGM concentrators in Rustenburg, South Africa was crushed, blended, and riffled. The ore fragments were then split into 1-kg samples by using a rotary splitter. A flotation slurry was then prepared by milling the 1-kg ore samples in synthetic plant water of the required ionic strength (see Table 4.1) at 66% solids. The 1% (w/v) SIBX solution was added to the mill at a dosage of 150 g/t. The collector solution was added to the mill because Wiese *et al.* [4] had shown this to be beneficial in sulfidic Cu–Ni–PGM ore flotation. This practice allows the collector to interact with the freshly created mineral particle surfaces. Milling was conducted in a laboratory scale stainless steel rod mill to achieve a grind of 60% passing 75 μm.

Table 4.1 Ion concentration, total dissolved solids (TDS), and ionic strength (I) in synthetically prepared plant water.

Water type	Ca^{2+} (mg/L)	Mg^{2+} (mg/L)	Na^{+} (mg/L)	Cl^{-} (mg/L)	SO_4^{2-} (mg/L)	NO_3^{-} (mg/L)	NO_2^{-} (mg/L)	CO_3^{2-} (mg/L)	TDS mg/L	I [M]
1 SPW	80	70	153	287	240	176	-	17	1023	0.0242
3 SPW	240	210	459	861	720	528	-	51	3069	0.0727
5 SPW	400	350	765	1435	1200	880	-	85	5115	0.1212
10 SPW	800	700	1530	2870	2400	1760	-	850	10230	0.2426

4.2.5 Batch Flotation

The slurry was transferred from the mill to a 3-L Barker flotation cell fitted with a variable speed drive and a manual pulp level controller. The volume of the slurry in the cell was made up to 3 L with the addition of the required synthetic plant water to produce a pulp containing 35% solids and the rest being water. The impeller speed was set to 1200 rpm.

CMC was added to the cell before the start of flotation at dosages of 100 and 500 g/t and conditioned for 2 min. Tests were also conducted in the absence of CMC, and 40 g/t of DOW 200 was added to the cell as the frother and conditioned for 1 min. This frother was supplied by Betachem. After all reagents were added and adequately conditioned, an air flow rate of 7 L/min was allowed into the cell. The froth height was allowed to build up to 2 cm. Both, the air flow rate and froth height were kept constant. Four concentrates were collected: C1 for 2 min, C2 for 4 min, C3 for 6 min, and C4 for 8 min by scraping the froth into a collecting pan every 15 s. Feed and tailing samples were taken before and after each flotation test. The amount of water recovered with each of the four concentrates was measured. All experimental tests were performed in duplicate, and reproducibility was found to be within 5% mass recovery in terms of the standard error.

4.2.6 Sample Assays/Analyses

After the collection of concentrates and tails, samples were filtered, dried, and weighed before analysis. This also allowed for the determination of the total solids and water recovered at the end of flotation for each experimental test. To determine the recoveries and grades of pyrrhotite from the recovered concentrate samples, copper and total nickel analyses were conducted using a Bruker X Ray Fluorescence (XRF) S4 Explorer Spectrometer. Sulfur analysis was carried out using a LECO DR423 sulfur analyzer. Assuming the stoichiometries for chalcopyrite and pentlandite to be $CuFeS_2$ and $(FeNi)_9S_8$ respectively, the recovery of iron sulfides (including pyrrhotite) was estimated from the copper, nickel, and sulfur values for each concentrate.

4.2.7 Pyrrhotite Sample Preparation

The pure mineral of key interest in this investigation was pyrrhotite. Pyrrhotite (1 kg) was supplied by Ward's Natural Science Establishment NY.

This was crushed manually using a hammer to 100 % passing 1000 μm sieve. The crushed sample was pulverized, and the fraction of +38-106 μm was used for automated contact timer, microflotation, and reagent adsorption studies. For the electrokinetic potential measurements (zeta potential), a fraction of the -38 μm grind size was used. Given that pyrrhotite readily oxidizes in air, samples thereof were purged with N_2 gas and stored in a refrigerator.

4.2.8 Bubble–Particle Attachment

An instrument known as an automated contact time apparatus (ACTA) was used to measure bubble-particle interactions. This instrument was developed at Aalto University in Finland [10, 18, 20]. The particle bed was made by firstly mixing 9 g of pyrrhotite with 100 mL of the particular synthetically prepared plant water under study. The slurry was conditioned for 1 min with SIBX, which was added at a dosage of 100 g/t. The slurry was allowed to settle upon, which the clear liquid or supernatant was pipetted out and filtered until about 2 cm of liquid remained above the settled particles. The filtrate was placed in the glass pool, after which the settled particles were pipetted into the pool for building the particle bed. Furthermore, to achieve a desired flat particle bed of 2 mm, an automated shovel was applied. Measurements were taken for each of the synthetic plant waters in Table 4.1. Tests were conducted at the natural pH, which was around 6.5. The temperature was recorded to be an average of 20°C. Experiments were performed in duplicate for each test condition.

4.2.9 Microflotation Tests

A microflotation cell that was developed by Bradshaw and O'Connor [21] to measure bubble loading in flotation was used to conduct the microflotation tests. Three grams of pyrrhotite were mixed with 50-mL of the synthetic plant water under study. The mixture was thereafter ultra-sonicated for 5 min to remove oxidation products from the mineral surface and prevent particle agglomeration. The slurry was then dispersed into the microflotation cell. A constant air flow of 7 mL/min was introduced to the microflotation cell, and the peristaltic pump circulating the pulp was set to 90 rpm. Two sets of tests were performed: the first set involved the presence of SIBX, while the second set involved the presence of CMC.

For the tests with a collector, a volume of 20 µL of the 1% (w/v) SIBX solution equivalent to 100 g/t dosage was added to the cell and conditioned for 1 min. For the tests with a depressant, a volume of 20 µL of the 1% (w/v) CMC solution equivalent to 100 g/t dosage was added to the cell and conditioned for 1 min.

Four concentrates were collected, after 2, 6, 12, and 20 min. The concentrates and tails were filtered and dried. The microflotation tests were performed in duplicate for each synthetic plant water. Like the tests with the ACTA, the slurry was at the natural pH of 6.5, and the temperature was recorded to be an average of 20°C for all experiments.

4.2.10 Zeta Potential Measurements

Firstly, synthetic plant water solutions of the water type under investigation were added into test beakers. The pH of the solution was not adjusted, as this needed to be around the natural pH of 6.5. Secondly, 0.0625 g of the prepared pyrrhotite sample with a particle size of -25 µm was added onto the test beakers. Thirdly, test beakers containing the suspension were placed on a magnetic stirrer for 15 min. The pH of the suspension was measured and corrected as necessary using dilute stock solutions of HCl and NaOH. Finally, 1 mL of the suspension was transferred into a Malvern Dip Cell and inserted into a Malvern Zetasizer for the determination of the zeta potential as a function of water quality. All zeta potential measurements were performed in triplicate to ensure reproducibility and reliability of the test results. It is important to note that for each condition, the third step was performed just before zeta potential determination in order to avoid any unnecessary time-dependent reactions with the inorganic electrolytes contained in the SPW that would blind the comparability of the results from one SPW to another.

4.2.11 Adsorption of Sodium Isobutyl Xanthate

The absorbances at 301 nm of five standard solutions were plotted against the known SIBX concentrations of these solutions to obtain a calibration curve. The Ultrospec 2100 UV-vis spectrophotometer (Biochrom, USA) was used for the adsorption studies.

In a conical flask, 3 g of pyrrhotite was mixed with 30 mL of the synthetic plant water under study. The pH of the slurry was then adjusted to 6.5 using the 1% (w/v) solutions of NaOH and HCl. The equivalent of 50 g/t of SIBX was added to the slurry. The conical flasks containing the slurry were covered with foil and inserted in an Ecobath shaker at 20°C and a shaking intensity of 141 rpm. After 15 min in the water bath, 10 mL of the slurry was drawn out with a plastic syringe. A 0.45-μm syringe filter was used to filter the slurry. The filtrate was then placed into a quartz cuvette, and the absorbance was measured at 301 nm. Measurements were corrected for the blanks for each of the SPW solutions. The equation of the calibration curve was used to calculate the unknown concentration of SIBX left in the solution and by mass balance, the concentration adsorbed on the mineral surface. This technique was applied to each of the SPW solutions with each condition tested in triplicates to minimize experimental error.

4.2.12 Adsorption of Sodium Carboxymethyl Cellulose

In order to investigate the extent to which CMC adsorbs onto pyrrhotite under varying ionic strengths of synthetic plant water, adsorption studies were conducted using the du Bois method (du Bois *et al.*, 1956) for the determination of the concentration of sugars. Nine grams of pyrrhotite was added to 90 mL of the particular synthetic plant water under investigation. The volume equivalent of 300 g/t of the 1% (w/v) CMC stock solution was added to the slurry. The slurry with CMC was then conditioned in an Ecobath for 3 min [23]. After conditioning, a 0.22-μm syringe filter was used to obtain 1 mL of the filtrate, subsequently, phenol and sulfuric acid were added as per the du Bois colorimetric method [24]. The absorbance was measured, and consequently, the residual concentration of CMC was determined. The experiments were conducted in triplicate for each synthetic plant water.

4.3 Results and Discussion

4.3.1 Batch Flotation of a Cu–Ni–PGM Ore: Effect of the Ionic Strength of SPW and CMC Dosage on the Behavior of Pyrrhotite

Figure 4.1 shows the relationship between pyrrhotite grade and recovery for all experimental conditions.

Figure 4.1 shows that at a depressant (CMC) dosage of 100 g/t, the recovery of pyrrhotite generally increased with an increase in the ionic strength of synthetic plant water (1, 3, and 10 SPW), except for 5 SPW,

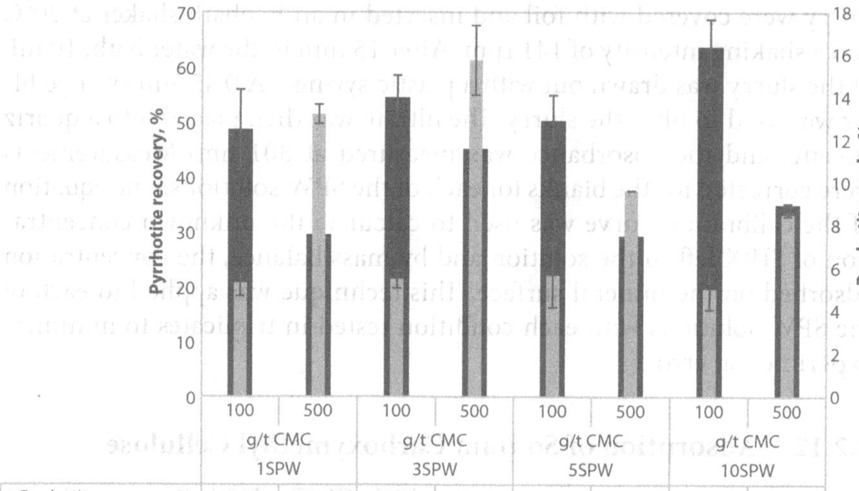

	1SPW 100 g/t CMC	1SPW 500 g/t CMC	3SPW 100 g/t CMC	3SPW 500 g/t CMC	5SPW 100 g/t CMC	5SPW 500 g/t CMC	10SPW 100 g/t CMC	10SPW 500 g/t CMC
Pyrrhotite recovery, %	48.58	29.46	54.40	45.12	50.20	29.21	63.68	34.95
Pyrrhotite grade, %	5.32	13.22	5.49	15.75	5.68	9.62	5.05	8.52

Figure 4.1 Total pyrrhotite recoveries and grades in increasing ionic strength of SPW and CMC dosages.

which resulted in a drop in pyrrhotite recovery compared to 3 SPW. A similar trend was seen for a CMC dosage of 500 g/t, except that further increases of the ionic strength of SPW from 5 SPW to 10 SPW, resulted in a further decrease in pyrrhotite recoveries. For a CMC dosage of 100 g/t, the recovery of pyrrhotite was in the order, 10 SPW>3 SPW>5 SPW>1 SPW. It is also interesting to note that for a given ionic strength of SPW, pyrrhotite recovery decreased upon the addition of CMC from 100 g/t to 500 g/t CMC. Figure 4.1 also shows that for a given SPW, pyrrhotite grades increased with an increase in CMC dosage. At 100 g/t CMC dosage, there seems to have been no discernable difference in pyrrhotite grade with a change in the ionic strength of SPW. However, for CMC dosage of 500 g/t, the concentrate grade with respect to pyrrhotite was in the following order 10 SPW<5 SPW<1 SPW<3 SPW.

In relating the effect of ionic strength and CMC on pyrrhotite to solids and water recoveries, Figure 4.2 shows the effect of CMC dosage and the ionic strength of SPW on the recovery of solids and water. The recovered solids are split into four categories, namely entrained non-sulfide gangue, naturally floating gangue, pyrrhotite, and other sulfides.

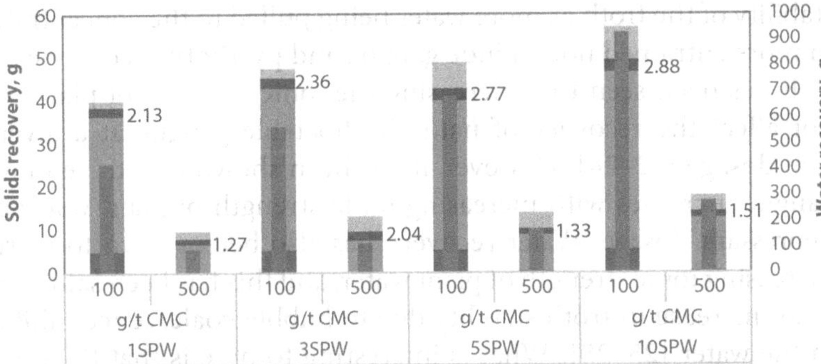

	1SPW 100 g/t CMC	1SPW 500 g/t CMC	3SPW 100 g/t CMC	3SPW 500 g/t CMC	5SPW 100 g/t CMC	5SPW 500 g/t CMC	10SPW 100 g/t CMC	10SPW 500 g/t CMC
Other Sulfides	1.36	1.73	2.13	3.45	5.93	3.62	7.40	3.56
Pyrrhotite	2.13	1.27	2.36	2.04	2.77	1.33	2.88	1.51
Entrained Gangue	31.27	6.63	37.57	7.44	34.65	8.45	41.07	12.26
Floating Gangue	5.13	0.00	5.23	0.04	5.30	0.40	5.45	0.44
Water	421.975	89.475	563.265	111.58	682.17	166.345	927.11	276.69

Floating Gangue · Entrained Gangue · Pyrrhotite · Other Sulfides · Water

Figure 4.2 Total solids (in split fractions of gangue and sulfides) and water recovery in increasing ionic strength of SPW and CMC dosages.

Figure 4.2 displays, for 100 g/t and 500 g/t CMC, a directly proportional trend between total solids and water recovery with increasing ionic strength of SPW as total solids and water recoveries increasing with an increasing ionic strength of SPW, while the increase in CMC dosage gave rise to a decrease in total solids and water recoveries. At a CMC dosage of 100 g/t, it can also be seen that for each ionic strength of SPW, an increase in CMC dosage from 100 g/t to 500 g/t gave rise to a decrease in pyrrhotite being pulled to the concentrate along with the decrease in water recoveries and entrained non-sulfide gangue. Increases in the ionic strength of SPW resulted in a slight increase in pyrrhotite recovered mass at a CMC dosage of 100 g/t. A similar trend was seen for the 500 g/t CMC dosage, except that 3 SPW was an outlier in that it reported a higher mass of pyrrhotite compared to all the other ionic strengths of SPW.

It must be noted that the recovered mass of pyrrhotite followed the trend seen with non-sulfide entrained gangue and water in relation to the changes in the ionic strength of plant water and CMC dosages. A more apparent trend of an increase in pyrrhotite recoveries with increasing ionic strength of plant water is seen in 100 g/t of CMC, thus corresponding well with the trends seen with the non-sulfide entrained gangue and water. This result is indicative of the role of the ionic strength of SPW and CMC dosage on the stability of the froth as more water being pulled to the concentrate resulted in more entrained non-sulfide gangue and pyrrhotite recoveries.

It has been seen that increasing the ionic strength of plant water does not affect the recovery of naturally floatable gangue at a given depressant dosage [22–24]. However, it has been shown that the entrainment of gangue increases with increasing ionic strength of plant water at a given depressant dosage. Water recovery has also been shown to increase with increasing ionic strength of plant water, and this has been said to be a result of an increase in froth stability due to bubble coalescence inhibiting ions in the water [25–28]. What is interesting to note is that the total sulfides recovery increased with an increase in the ionic strength of SPW, although previously chalcopyrite and pentlandite recoveries from a similar ore were reportedly unaffected by increases in the ionic strength of SPW [22, 23, 29, 30]. This, therefore, suggests that the influence of ore type is worth considering in future studies since the mineralogy of the ore may very well affect interactions occurring in the pulp phase.

Pyrrhotite recoveries showed an increase with an increase in the ionic strength of SPW at a given CMC dosage. For 5 and 10 SPW, pyrrhotite recoveries took a dip in comparison to the trend of an increase from 1 SPW>3 SPW, though 10 SPW still reported higher recoveries compared to 5 SPW. This phenomenon with 5 SPW suggests that there is a threshold in terms of ion concentration beyond which interactions at the pyrrhotite–water interface may either result in a decrease or increase in pyrrhotite recoveries. This further supports claims made by Dzingai *et al.* [31, 32] who showed that there exist specific ion threshold concentrations beyond which froth stability, mineral recoveries, and grades are affected negatively.

4.3.2 Microflotation of Pyrrhotite in Increasing Ionic Strength of SPW and CMC Dosage

Figure 4.3 shows the microflotation data for pyrrhotite in increasing ionic strength of SPW in systems without CMC and those with CMC at a dosage of 300 g/t.

Figure 4.3 shows the microflotation data for pyrrhotite: (a) presents the observations in the absence of CMC; and (b) presents the observations in the presence of CMC at a dosage of 300 g/t. Figure 4.3 shows an increase in the pyrrhotite recovery with increasing ionic strength of SPW. In the absence of CMC, at the end of flotation, the recoveries of pyrrhotite were 32.95%, 38.25%, and 48.87% for 1 SPW, 5 SPW, and 10SPW, respectively. In the presence of CMC (Figure 4.3(b)), the recoveries of pyrrhotite at the

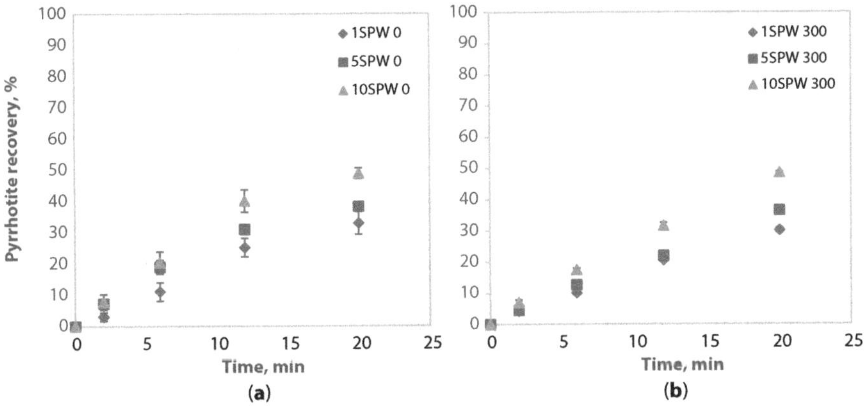

Figure 4.3 Pyrrhotite recoveries in increasing ionic strength of SPW in (a) the absence of CMC and in (b) the presence of CMC at a dosage of 300 g/t.

end of flotation were 30.09%, 36.47%, and 48.54% for 1 SPW, 5 SPW, and 10 SPW, respectively. Both trends show that increasing the ionic strength of SPW resulted in an increase in pyrrhotite recoveries. These results agree with the batch flotation results. Furthermore, the addition of CMC to the system resulted in a decrease in pyrrhotite recoveries. This further supports the batch flotation results on the recoveries of pyrrhotite with CMC. The slight decrease seen in pyrrhotite recoveries in the presence of CMC suggests that the conditions of SPW presented an oxidizing environment for the mineral. This, therefore, created a conducive environment for CMC adsorption onto pyrrhotite, which further decreased the recovery of the mineral. CMC adsorption results are presented in subsequent sections in this chapter. The increase in pyrrhotite recoveries with increasing ionic strength of SPW could suggest one or two mechanisms or a combination of both: (1) the bubble–particle attachment of pyrrhotite may have improved in increasing ionic strength of SPW and (2) the recovery of pyrrhotite increased in increasing ionic strength of SPW as a result of an increase in entrainment, and/or as can be directly correlated with the increase in the amount of entrained non-sulfide gangue seen in Figure 4.2.

4.3.3 Bubble–Particle Attachment of Pyrrhotite in Increasing Ionic Strength of SPW

Figure 4.4 shows the results of the attachment probability of pyrrhotite particles in increasing ionic strength of SPW in the presence of SIBX using the Automated Contact Time Apparatus.

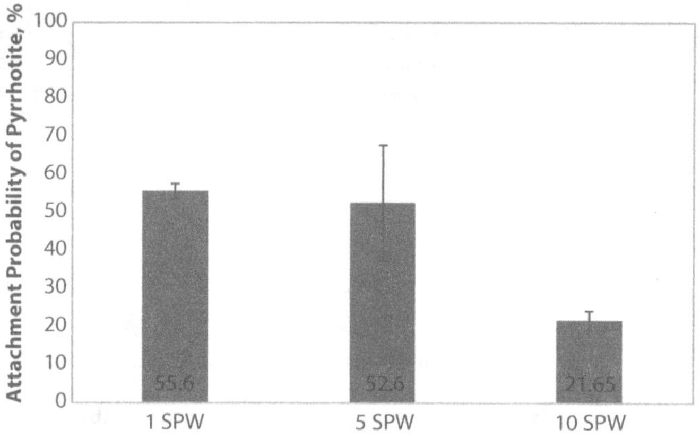

Figure 4.4 Attachment probability of pyrrhotite under increasing ionic strength of SPW.

Figure 4.4 shows a decrease in the attachment probability of pyrrhotite particles with increasing ionic strength of SPW. This decrease in attachment probability could be caused by changes on the surface chemistry of pyrrhotite, affecting the hydrophobicity of the pyrrhotite particles in process water of a degrading quality. For this reason, the adsorption of SIBX onto pyrrhotite in process water of a degrading quality was studied.

It must be stated that the lower attachment probability at high ionic strength is not in line with increases in pyrrhotite recoveries seen in Figures 4.1, 4.2, and 4.3. This, therefore, suggests that another mechanism to the recovery of pyrrhotite in the batch flotation test work may be responsible for the increased recoveries of pyrrhotite, as seen in Figures 4.1 and 4.2, that mechanism being entrainment beyond any speculations of an increase in the hydrophobization and floatability of pyrrhotite in increasing ionic strength of SPW.

4.3.4 Adsorption of Sodium Isobutyl Xanthate onto Pyrrhotite in Increasing Ionic Strength of SPW

Figure 4.5 shows the results of the adsorption studies of SIBX onto pyrrhotite particles in increasing ionic strength of SPW. These results are presented in the form of residual SIBX concentration, that is the concentration of SIBX remaining in solution, after some SIBX had adsorbed onto the mineral surface.

From Figure 4.5, it can be seen that as the ionic strength of SPW increases, the concentration of SIBX left in solution increases. Therefore, the higher

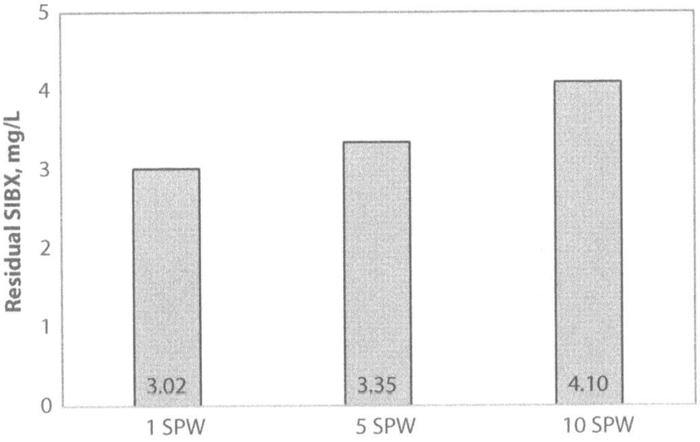

Figure 4.5 Residual sodium isobutyl xanthate under increasing ionic strength of SPW.

the ionic strength of the SPW, the higher the residual concentration of SIBX, thus signifying that in increasing ionic strength of SPW, there was a decrease in the adsorption of SIBX onto the pyrrhotite surface. Poor adsorption of SIBX onto mineral surfaces has a negative effect on the hydrophobicity of the mineral particles; this could therefore explain the decrease in attachment probability of the pyrrhotite particles to air bubbles under degrading water quality, as seen in Figure 4.4.

It would also be interesting to understand the effect of process water of degrading quality on the adsorption of the depressant onto the pyrrhotite surface, and see how that relates to the floatability of pyrrhotite, as seen in Figures 4.1, 4.2, and 4.3(b).

4.3.5 Adsorption of Carboxy Methyl Cellulose onto Pyrrhotite in Increasing Ionic Strength of SPW

Figure 4.6 shows the results of the adsorption studies of CMC onto pyrrhotite particles in increasing ionic strength of SPW. It can be seen that the higher the ionic strength of synthetic plant water, the lower the residual concentration of CMC. Therefore, the adsorption of CMC onto the pyrrhotite surface increased with degrading water quality.

These results along with the results presented in Figure 4.5 support the postulation that in increasing ionic strength of SPW, the hydrophilicity of pyrrhotite particles is induced, and hence the decrease in the attachment probability with increasing ionic strength of SPW (see Figure 4.4).

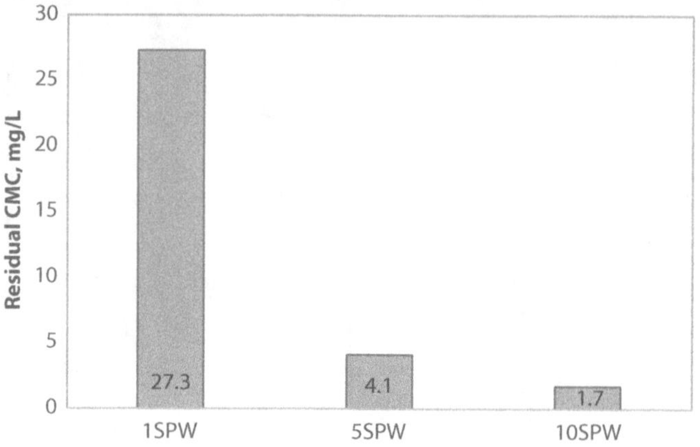

Figure 4.6 Residual sodium carboxy methyl cellulose under increasing ionic strength of SPW.

4.3.6 Zeta Potential of Pyrrhotite in Increasing Ionic Strength of SPW

Figure 4.7 shows the zeta potential of pyrrhotite particles under increasing ionic strength of synthetic plant water at the natural pH of the pyrrhotite-SPW slurry. These results are all in the absence of CMC or SIBX. It is clear that the zeta potential became less negative with increasing ionic

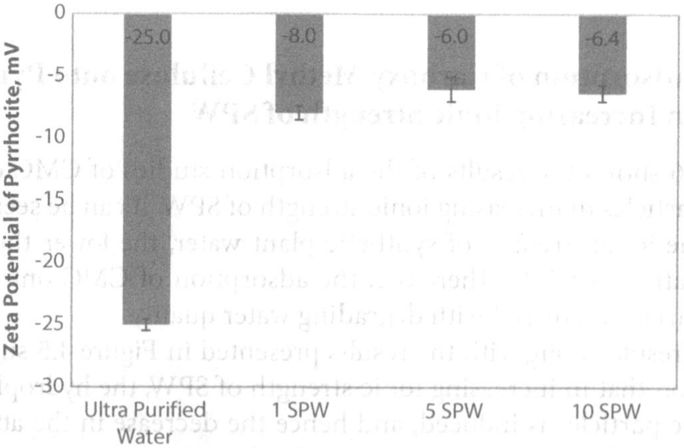

Figure 4.7 Zeta potential of pyrrhotite under increasing ionic strength of SPW.

strength of plant water. Therefore, it can be said that as the ionic strength of the synthetic plant water increases, the magnitude of the zeta potential of pyrrhotite decreases. A less negative zeta potential with higher ionic strength implies an increase in the adsorption of metal cations onto the mineral surface which passivate the mineral surface. This passivation of the pyrrhotite mineral surface is conducive for CMC adsorption, resulting in the increased hydrophilicity of the pyrrhotite particles, and thus decreasing the floatability of pyrrhotite.

These results further support the theory of an induced acid-base interaction between a passivated pyrrhotite mineral surface and the highly negatively charged CMC ligand [33–35]. This has been proposed as the mechanism through which the CMC adsorption onto pyrrhotite is increased in the presence of cations, such as those present in the of SPW used in this study. This further supports the results in Figure 4.6 showing that increasing the *I of SPW* gave rise to an increase in CMC adsorption (decrease in residual CMC), and hence the subsequent depression of pyrrhotite seen through the lower recoveries with synthetic plant water of higher ionic strength in Figures 4.1 and 4.2.

The passivation of the pyrrhotite mineral surface, however, seems to have hindered the adsorption of SIBX on the mineral surface as shown by the increase in residual SIBX in solution with synthetic plant waters of increased ionic strength. Thus, indicating that the adsorption of the divalent metal cations and hydroxo species onto the mineral surface reduces the sites available on the pyrrhotite surface for the adsorption of SIBX; and consequently, reducing the hydrophobicity of pyrrhotite, which can also be seen by the lower attachment probability under higher ionic strengths of SPW.

4.3.7 Concluding Discussion

The results from this investigation have shown that entrained gangue, floating gangue, and water recoveries increased with an increase in the ionic strength of plant water and decreased with an increase in CMC dosages. It must be mentioned that the resulting increase in total solids and water recoveries in increasing ionic strength of plant water can be attributed to an increase in froth stability [22, 29, 34, 35]. Observations from this work on entrained gangue and floating gangue correlated well with the water recovery trend and were therefore in agreement with the fact that increases in the ionic strength of synthetic plant water, increase the stability of the froth whereas increases in CMC dosages are known to decrease the stability of the froth [36, 37]. Pyrrhotite recoveries, however, were not easily comparable to the trends seen on other sulfides and floatable gangue, but instead followed the same trends as those of the entrained gangue and water recoveries, and thus could be correlated with the impact that the ionic strength of plant water and CMC dosage have on the stability of the froth. It must also be noted that pyrrhotite flotation is known to be particularly complicated and highly dependent on oxidation potential and pH. The trend seen on entrained gangue can thus be attributed to the rise in froth stability, as well as the activation of gangue particles, which occurs in the presence of inorganic electrolytes, thus causing surface hydration of naturally hydrophobic minerals to decrease [7]. The naturally hydrophobic pyrrhotite, a metallic conductor highly prone to oxide, floats due to the production of hydrophobic elemental sulfur species [38], but the same oxidation could result in ferric hydroxide sites on the pyrrhotite surface resulting in a hydrophilic mineral surface, therefore prone to depression along with non-sulfide gangue minerals.

Yoon and Sabey [39] suggested that an ionic strength increase results in a further increase in hydrophobicity. This could explain the increase in the recovery of pyrrhotite with increasing ionic strength of plant water for the batch flotation and microflotation results. This phenomenon can be attributed to inorganic electrolytes, which reduce repulsive forces between bubbles and pyrrhotite particles by decreasing the zeta potential as was shown in October *et al.* [10, 11]. This expected increase in hydrophobicity with increasing ionic strength of plant water was, however, not observed in the attachment time results with SIBX. Therefore, in this pyrrhotite system, the effect of the electrical double layer and the subsequent reduction in repulsive forces between bubble and particle may be concealed by the effects of oxidation and the formation of oxyhydroxo species on the SIBX adsorption.

From the results of this investigation, it can be deduced that the behavior and floatability of pyrrhotite in polymetallic ores, when changes in the ionic strength of SPW and CMC dosages are implemented, cannot be simply correlated to the behavior of other sulfides, such as chalcopyrite, pentlandite, and non-sulfide naturally floatable gangue. However, water recovery and entrained gangue, which can be directly correlated with the impact on the stability of the froth, are key parameters to monitor and can predict the trend that pyrrhotite particles will follow.

Thus, it can be said that the ionic strength of plant water and CMC dosage electrolytes are important variables to consider when tracking the behavior of pyrrhotite from BMS-PGM ore flotation. With the activity of the CMC depressant being affected by interactions between gangue minerals and pyrrhotite in the pulp, it is even more important to examine the behavior seen from the results of this work in terms of decoupling pyrrhotite particles reporting to the concentrate through true flotation and those that do via entrainment.

4.4 Conclusions

The findings of this work showed that an increase in ionic strength results in an increase in pyrrhotite recoveries. This finding corresponds well with the general trend of an increase in the recoveries of non-sulfide entrained gangue and water recoveries. Increasing the depressant dosage of CMC 100 g/t to 500 g/t resulted in a decrease in pyrrhotite recoveries owing to two possible mechanisms: (1) increase in the hydrophilicity of pyrrhotite particles as a result of an increase in CMC adsorption onto pyrrhotite and (2) decrease in the amount of water recoveries, thus decreasing froth stability and mineral particle entrainment.

Attachment time tests with pyrrhotite and SIBX showed a decrease in pyrrhotite bubble–particle attachment with increasing ionic strength of plant water. This trend was supported by the decrease in SIBX adsorption and increase in CMC adsorption in increasing ionic strength of SPW; further supporting a postulation that increases in the ionic strength of SPW result in a decrease in the hydrophobicity of pyrrhotite particles.

Overall, in process water of a degrading quality, pyrrhotite is passivated due to the presence of polyvalent cations, less collector adsorbs on the mineral surface, the surface chemistry is conducive for CMC adsorption, thus reducing its hydrophobicity, bubble–particle attachment and resulting in lower floatability. However, in a flotation system, pyrrhotite recoveries may increase as a result of an increase in entrainment of particles.

Acknowledgments

This work was supported in part by the National Research Foundation (NRF) of South Africa (SA) (Grant numbers 99262 and 103641) and the Academy of Finland Mineral Resources and Material Substitution MISU program (Finland) – Protocol development for evaluation of water-saving alternatives in minerals processing – "Bridging North to South" project. Any opinions, findings, conclusions, or recommendations expressed in any publication generated by NRF-supported research is that of the authors and the NRF accepts no liability whatsoever in this regard. This work has also been part of the South African Minerals to Metals Research Institute (SAMMRI) Project No. S1612 funded by the Department of Science and Technology, SA. Financial and technical support from the members of the UCT CMR's Reagent Research Group is also acknowledged. The authors would like to acknowledge support from the University of Cape Town and members of the Reagent Research Group (RRG) in the Centre for Minerals Research (CMR).

References

1. Hochreiter, R.C., Kennedy, D.C., Muir, W., Woods, A.I., Platinum in South Africa (metal review series no.3). *J. South. Afr. Inst. Min. Metall.*, 85, 165–185, 1985.
2. Allison, S.A. and O'Connor, C.T., An investigation into the flotation behaviour of pyrrhotite. *Miner. Eng.*, 98, 3-4, 202–207, 2011.
3. Becker, M., Harris, P.J., Wiese, J.G., Bradshaw, D.J., Mineralogical characterisation of naturally floatable gangue in merensky reef ore flotation. *Int. J. Miner. Process.*, 246–255, 2009.
4. Wiese, J., Harris, P., Bradshaw, D., The influence of the reagent suite on the flotation of ores from the merensky reef. *Miner. Eng.*, 18, 189–198, 2005.
5. Corin, K.C., Reddy, A., Miyen, L., Wiese, J.G., Harris, P.J., The Effect of Ionic Strength of Plant Water on Valuable Mineral and Gangue Recovery in a Platinum Bearing Ore from the Merensky Reef. *Miner. Eng.*, 24, 131–137, 2011.
6. Rao, S.R. and Finch, J.A.A., Review of Water Re-Use in Flotation. *Miner. Eng.*, 2, 65–85, 1989.
7. Manono, M.S., *Investigating Electrolyte-Reagent-Mineral Interactions in Response to Water Quality Challenges in the Flotation of a PGM Ore*, PhD Thesis, University of Cape Town, 2018.
8. Manono, M., Corin, K., Wiese, J., Perspectives from literature on the influence of inorganic electrolytes present in plant water on flotation performance. 54, 1191–1214, 2018.
9. Hodgson, M. and Agar, G.E., Electrochemical investigations into the flotation chemistry of pentlandite and pyrrhotite: Process water and xanthate interactions. *Can. Metall. Q.*, 28, 189–198, 1989.

10. October, L., Corin, K., Schreithofer, N., Manono, M., Wiese, J., Water quality effects on bubble-particle attachment of pyrrhotite. *Miner. Eng.*, 131, 2019.
11. October, L.L., Manono, M.S., Corin, K.C., Schreithofer, N., Wiese, J.G., The influence of specific ions and oxyhydroxo species in plant water on the bubble-particle attachment of pyrrhotite. *ACS Omega*, 6, 28496–28506, 2021.
12. Becker, M., de Villiers, J., Bradshaw, D., The flotation of magnetic and non-magnetic pyrrhotite from selected nickel ore deposits. *Miner. Eng.*, 23, 1045–1052, 2010.
13. Vaughan, D.J., Becker, U., Wright, K., Sulphide mineral surfaces: Theory and experiment. *Int. J. Miner. Process.*, 51, 1–14, 1997.
14. Brough, C.P., Bradshaw, D.J., Becker, M., A comparison of the flotation behaviour and the effect of copper activation on three reef types from the merensky reef at Northam. *Miner. Eng.*, 23, 846–854, 2010.
15. Ekmekçi, Z., Becker, M., Tekes, E.B., Bradshaw, D., The relationship between the electrochemical, mineralogical and flotation characteristics of pyrrhotite samples from different ni ores. *J. Electroanal. Chem.*, 647, 133–143, 2010.
16. Ekmekçi, Z., Becker, M., Tekes, E.B., Bradshaw, D., An impedance study of the adsorption of CuSO4 and SIBX on pyrrhotite samples of different provenances. *Miner. Eng.*, 23, 903–907, 2010.
17. Chimbganda, T., Becker, M., Broadhurst, J.L., Harrison, S.T.L., Franzidis, J.P., A comparison of pyrrhotite rejection and passivation in two nickel ores. *Miner. Eng.*, 46–47, 38–44, 2013.
18. Aspiala, M., Schreithofer, N., Serna-Guerrero, R., Automated contact time apparatus and measurement procedure for bubble-particle interaction analysis. *Miner. Eng.*, 121, 77–82, 2018.
19. Javor, Z., Aspiala, M., Schreithofer, N., Heiskanen, K., Serna, R., Development of a new attachment timer for predicting the change in ore floatability, in: *Proceedings of the Proceedings of the XXVIII International Mineral Processing Congress (IMPC 2016)*, Québec City, Canadian Institute of Mining, Metallurgy and Petroleum, September 11-15, 2016, 2016.
20. Jávor, Z., Schreithofer, N., Heiskanen, K., Multi-scale analysis of the effect of surfactants on bubble properties. *Miner. Eng.*, 99, 170–178, 2016.
21. Bradshaw, D.J. and Connor, C.T., Measurement of the sub-process of bubble loading in flotation. *Miner. Eng.*, 9, 443–448, 1996.
22. Corin, K.C., Reddy, A., Miyen, L., Wiese, J.G., Harris, P.J., The effect of ionic strength of plant water on valuable mineral and gangue recovery in a platinum bearing ore from the merensky reef. *Miner. Eng.*, 24, 131–137, 2011.
23. Moimane, T.M., Corin, K.C., Wiese, J.G., The effect of varying pulp reagent chemistry on the flotation performance of a South African PGM ore. *Miner. Eng.*, 95, 155–160, 2016.

24. Manono, M., Corin, K., Wiese, J., Water quality effects on a sulfidic pgm ore: Implications for froth stability and gangue management. 54, 1253–1265, 2018.
25. Lessard, R.R. and Zieminski, A.S., Bubble coalescence and gas transfer in aqueous electrolytic solutions. *Ind. Eng. Chem. Fundam.*, 10, 260–269, 1971.
26. Zieminski, S.A. and Whittemore, R.C., Behaviour of gas bubbles in aqueous electrolyte solutions. *Chem. Eng. Sci.*, 26, 509–520, 1971.
27. Craig, V.S.J., Ninham, B.W., Pashley, R.M., The effect of electrolytes on bubble coalescence in water. *J. Phys. Chem.*, 97, 10192–10197, 1993.
28. Manono, M.S., Corin, K.C., Wiese, J.G., The effect of ionic strength of plant water on foam stability: a 2-phase flotation study. *Miner. Eng.*, 40, 42–47, 2013.
29. Corin, K.C. and Wiese, J.G., Investigating froth stability: A comparative study of ionic strength and frother dosage. *Miner. Eng.*, 66, 130–134, 2014.
30. Moimane, T.M., Corin, K.C., Wiese, J.G., Investigation of the interactive effects of the reagent suite in froth flotation of a merensky ore. *Miner. Eng.*, 96–97, 39–45.
31. Dzingai, M., Manono, M., Corin, K., Simulating the effect of water recirculation on flotation through ion-spiking: Effect of Ca2+ and Mg2+. *Minerals*, 10, 1033, 2020.
32. Dzingai, M., Manono, M.S., Corin, K.C., Probing the effect of water recycling on flotation through anion spiking using a low-grade Cu–Ni–PGM ore: The effect of NO3–, SO42– and S2O32–. *Minerals*, 11, 340, 2021.
33. Laskowski, J.S., Liu, Q., Bolin, N.J., Polysaccharides in flotation of sulphides. part i. adsorption of polysaccharides onto mineral surfaces. *Int. J. Miner. Process.*, 33, 223–234, 1991.
34. Liu, Q., Zhang, Y., Laskowski, J.S., The adsorption of polysaccharides onto mineral surfaces: An acid/base interaction. *Int. J. Miner. Process.*, 60, 229–245, 2000.
35. Laskowski, J.S., Liu, Q., O'Connor, C.T., Current understanding of the mechanism of polysaccharide adsorption at the mineral/aqueous solution interface. *Int. J. Miner. Process.*, 84, 59–68, 2007.
36. Manono, M.S., Corin, K.C., Wiese, J.G., An investigation into the effect of various ions and their ionic strength on the flotation performance of a platinum bearing ore from the merensky reef. *Miner. Eng.*, 36–38, 2012.
37. Manono, M.S., Corin, K.C., Wiese, J.G., The effect of ionic strength of plant water on foam stability: A 2-phase flotation study. *Miner. Eng.*, 40, 2013.
38. Schreithofer, N., Wiese, J., McFadzean, B., Harris, P., Heiskanen, K., OConnor, C., Frother-depressant interactions in two and three phase systems. *Int. J. Miner. Process.*, 100, 33–40, 2011.
39. Wiese, J., Harris, P., Bradshaw, D., The Response of Sulphide and Gangue Minerals in Selected Merensky Ores to Increased Depressant Dosages. *Miner. Eng.*, 20, 986–995, 2007.

5

Simulated Short Cycle Water Recirculation on the Flotation Performance of a UG2 Cu–Ni–PGM Ore

Malibongwe S. Manono*, Michael S. Ngau and Kirsten C. Corin

Centre for Minerals Research, Department of Chemical Engineering, University of Cape Town, Rondebosch, South Africa

Abstract

Water recirculation, as a water management strategy, has developed into the standard process for water use in mineral beneficiation plants. This has been promoted as a more sustainable strategy as it ensures a reduction of used water effluents and freshwater withdrawals by the plants. Water recirculation, however, results in a build-up of ions present in the solution, which may affect flotation performance. The impact of the ions accumulation, resulting from water recirculation, is usually studied through the synthetic build-up of ions concentration, and not through process water recovery as proposed in this project. In this study, the impact of short water recirculation on flotation performance is investigated through the monitoring of flotation performance indicators. The experimental methodology developed in this study focused on (1) investigating the impact of depressant dosage; (2) investigating the impact of point of addition of recirculated water (i.e., to the mill or the float cell.); (3) investigating the impact that water recirculation has on ionic concentration of Ca^{2+}, Mg^{2+}, SO_4^{2-}, and Cl^-; and (4) investigating the impact that water recirculation has on the solids and water recoveries, and the grade and recovery of sulfides. The techniques utilized in this study were as follows: (1) elemental analyses using X-ray fluorescence (XRF) and (2) colorimetric analysis of tailings water samples for the presence of Ca^{2+}, Mg^{2+}, SO_4^{2-}, and Cl^-; and (3) mass balance analysis of the solids and water recoveries from the batch scale flotation of the UG2 ore. The addition of a depressant between 0 and 300 g/t led to a decrease in solids

**Corresponding author*: malibongwe.manono@uct.ac.za

Elvis Fosso-Kankeu, Bhekie B. Mamba and Antoine F. Mulaba-Bafubiandi (eds.) Recovery of Values from Low-Grade and Complex Minerals: Development of Sustainable Processes, (111–144) © 2024 Scrivener Publishing LLC

recovery, a three-fold increase in copper grade, and a two-fold increase in nickel grade. Water recirculation led to an increase in the ionic concentration of the pulp and froth, improving the froth stability and enhancing entrainment. In comparison to synthetically prepared fresh plant water, floating with recirculated water led to an increase in solids and water recoveries. However, it reduced the grades of recovered sulfides due to an increase in gangue recovery. Short cycle water recirculation led to slightly higher grades of copper and a 2% increase in nickel recoveries in recirculated water to the float cell, as compared to the mill. Similarly, float cell recirculated waters also had higher Ca^{2+}, Mg^{2+}, and SO_4^{2-} ion concentrations.

***Keywords*:** Flotation, flotation reagents, mineral recovery, water recirculation

5.1 Introduction

Although mining in South Africa contributed 5% of the water consumption in 2012, as reported by Statistics South Africa [1], mining companies reported negative impacts related to water in the last decade in the 2010 Water Disclosure Project run by the National Water Initiative [2]. The scarcity of fresh water in South African rivers, exaggerated by increased freshwater withdrawals by a growing population, climate change reducing green and blue water resources, and legislative action for stringent environmental restrictions, has motivated industrial innovation in the usage of fresh water and handling of wastewaters [3, 4]. Alternative water sources, such as treated sewage water and groundwater, and water management strategies, such as internal water recycling (long and short), reuse, and reduction are being proposed, and are utilized in the mining operations [5]. These strategies may inadvertently affect water quality and may, in turn, affect flotation performance.

This has motivated a significant body of research into the impact of water quality on froth flotation performance [3, 6–18]. These studies note significant deviation in flotation performance of similar ores using different water qualities. However, modern concentrator design and optimization utilizing metallurgical batch flotation and pilot-scale tests use fresh tap water at the laboratory site, leading to a gap between laboratory results and actual plant performance [19–21].

The variable water sources differ significantly in chemistry and may inadvertently affect flotation chemistry. This owes to the influence of different species present in solution (as some may be biotic whereas others may be abiotic), total suspended solids (TSS), and residual reagents, which may affect water chemistry and reaction pathways, and inadvertently affect mineral particle surface, air bubbles, and aqueous solution [6, 7, 9, 22, 23].

These changes affect sub-processes of flotation such as bubble-to-mineral particle collision, attachment, and flotation of stable particle-bubble aggregates, which are buoyant [24]. Water constituents can also react with reagents added to flotation circuits, changing the reagents' properties and impacting the flotation performance [14, 25].

Literature indicates that dissolved species in solution, and water quality, pH and DO, can affect flotation performance with both beneficial and detrimental effects in their applicable ranges [3, 6, 13, 20, 26–32]. Meanwhile, water quality variation in flotation is attributed to both internal and external factors of the concentrator. For internal factors, these include the type of ore being processed, reagents added and water internal reuse [33]. On the other hand, external factors can be divided as being either external to the concentrator or external to the site. External to the concentrator factors are defined as factors that are within the site boundary including raw water streams and water reuse from tailing storage facilities [33]. Of key importance is the impact of internal factors on the water quality as these factors are amplified by the accumulation of species through short water recirculation, leading to an increase in the total dissolved solids (TDS) content [9].

Water recirculation has been adopted as a water management strategy in the mineral processing industry accounting for between 60% and 65% of water consumed in some plants in 2015 [34]. The benefit is that water recirculation leads to stabilized flotation rates, improves flotation kinetics from residual reagents, and prevents environmental pollution through mine water discharge or effluent. It is asserted as an attractive alternative to meet the zero-wastewater targets from operations and minimizes freshwater withdrawal [6, 32, 35, 36]. Forssberg and Hallin [7] stated that a reduction of 50% residual reagent to the feed is sometimes required when using recirculated water. The effect of water recirculation is sometimes delayed because of the time delay from the tailings storage facilities to the plant [7]. It has also been suggested that water recirculation leads to a decrease in reagent efficiency as a result of contaminants, an increase in suspended solids, and dissolved ion concentration within the recycled water, residual reagents, and secondary effects of increased pollutant levels in the recycled water such as chemical and microbiological oxidation products [10, 11].

Of note, xanthate in sulfide flotation oxidizes and decomposes to produce dixanthogen, which is known to induce non-selective flotation [19]. However, this is not always observed [9]. Of key interest are organic reagents due to their chemical oxygen demand (COD) levels, which have motivated the use of different technologies to remove them from the water [3, 11, 37]. It has been shown that water re-circulation leads to an increase in water ionic strength, leading to the compression of the electrical double layer, reducing the electrostatic repulsion affecting the air-water interface, reducing the rate of coalescence, and reducing bubble rupture, thereby making the froth more stable [3, 6, 32, 38]. A more stable froth may, in turn, result in increased recoveries at the expense of the concentrate grade owing to an increase in the entrainment of gangue. It has also been said that the accumulation of fine gangue and clay minerals in recirculated water could cause the formation of slime coating on the surface of valuable minerals, e.g., slime coatings on pentlandite may modify the mineral surface charge, thereby impeding its flotation [14, 25].

Water recirculation negatively impacts selectivity when two or more concentrates are produced in a sequential circuit as it decreases the selectivity of the process in that some residual reagents may be present in the recirculate water, which may be detrimental to the flotation of some minerals, e.g., residual xanthate in recycled process water from the nickel flotation circuit is detrimental when returned to the copper flotation circuit in a copper–nickel sequential circuit [19, 36]. Water recirculation may also accumulate heat and nutrients, favoring the development of bacteria and organic species; these could act as surface activators, dispersants, or flocculants and interfere with flotation [7, 8, 11, 17].

It is worth mentioning that water management is not a common feature in plant design and process selection. Evaluating changes at a plant scale is often costly because of the long transition time between changes and effect actualization, limiting statistical analysis of the results [39]. The impact of a build-up of species in solution has been well studied [6, 13, 18, 38, 40–45]. However, laboratory investigations dedicated to the study of variation in process water quality due to closed water circuits are novel, with Manenzhe *et al.* [20] pioneering simulated water recovery and dissolution and test protocols, respectively, to fill this gap in the literature. Prediction of the quality of mining wastewater has been considered in the literature [46–48]. However, these methods possess disadvantages since thermodynamic predictions may not be effective in cases when kinetic reactions govern the system [47]. In short water recycling systems, the kinetics might govern performance since the water quality variation can be expeditious [49, 50].

Therefore, this study is aimed at simulating short water recirculation to allow for the analysis of the variation of water quality over time, owing to the impact of internal operational factors within a closed recycle water circuit and applying the results to predict the impact on flotation performance. This information will be used in comparison to simulated recycled water quality research work done.

5.2 Materials and Methods

The experimental procedure utilizes synthetic plant water recipe developed by the Centre for Minerals Research (CMR) to mimic typical water quality determined at several platinum group mineral (PGM) concentrators across the Bushveld Ingenuous Complex [51]. It is assumed that quality variations can be mimicked if the pathway of the ore in the plant can be mimicked. It is also assumed that if the water quality could be predicted, then, with some degree of uncertainty, its effect on the flotation and plant maintenance could be predicted. It is also assumed that it could be beneficial as complementary to the traditional synthetic plant water approach for studying the effect of recycling water on flotation performance.

The experimental program was divided into two phases. Phase 1 constituted a series of baseline tests to establish the relationship between depressant dosage and solids and gangue recovery so as to establish the entrainment factor (i.e., the degree of the entrainment of gangue). Phase 2 was conducted to evaluate the impact of water recirculation on flotation performance, as well as point of addition of recycled water. An outline of the phases and design of the experiment is illustrated in Figure 5.1. One factor at a time design was used for the experimental conditions tested on the UG2 ore. All measurements were conducted in triplicate to allow for

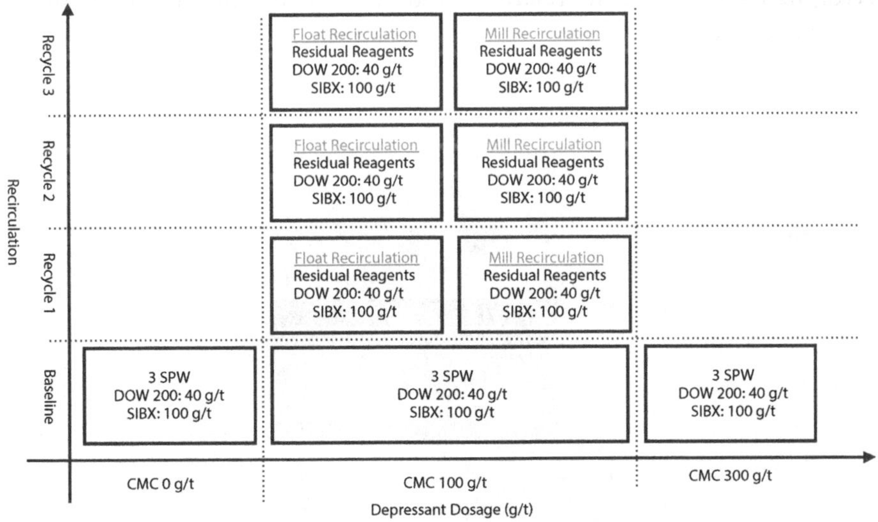

Figure 5.1 One factor at a time design for the experimental conditions tested on the UG2 ore.

statistical analysis of the results. Subsequent error analysis was found to be within 5% for solids and 10% for water recovered.

5.2.1 UG2 Ore Mineralogy

The UG2 ore sample was obtained from Impala Platinum, situated within the Bushveld Igneous Complex (BIC) in South Africa. The ore is a low-grade Cu–Ni PGM ore, and is the second most mined source of Base Metal Sulfides (BMS) in the Bushveld Complex. The sulfide minerals present are <1% of the ore and are composed of chalcopyrite (roughly 10% of all sulfides), pentlandite (roughly 35% of all sulfides), and pyrrhotite (roughly 50% of all sulfides) [52, 53]. The valuable PGMs present in the ore (~0.1%) are roughly 45% platinum (Pt), 10% rhodium (Rh), 25% palladium (Pd), 15% ruthenium (Ru), iridium (Ir), and osmium (Os), and together with gold (Au) and silver (Ag) are known as precious metals [53]. Gangue minerals of interest in the UG2 ore are chromite (70%), pyroxene (15 – 20%), feldspar (3 – 5%), and talc (1 – 3%), and are considered owing to their possible impact on both grade and entrainment in the flotation concentrates [54]. The bulk UG2 ore was supplied in 2019 by Impala Platinum mines in 220-L clamp-sealed drums as blended and crushed in four 15-kg ore bags. The ore sample was riffled and split into 1.3-kg portions using a vibrating rotary splitter. This approach produces a consistent composition between the test charges, as the riffler reduced the group and segregation error, hence reducing the relative standard deviation of the sample mean pay-metal grade.

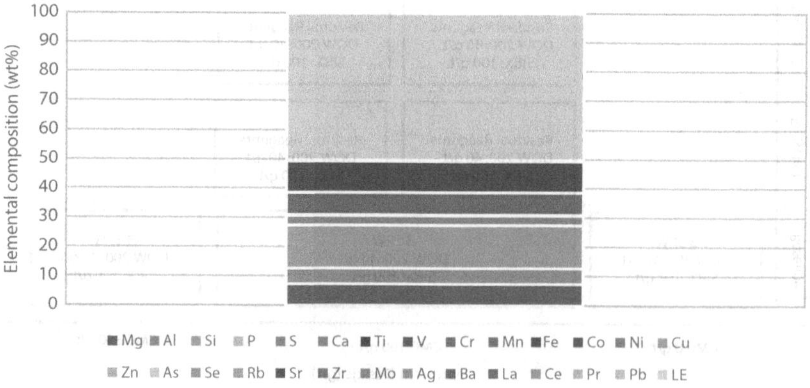

Figure 5.2 Individual elemental composition of the UG2 ore feed analyzed using X-ray fluorescence (XRF).

Table 5.1 Mean calculated feed values for the UG2 ore received in 2019.

	Copper, wt%	Nickel, wt%
UG-ore sample	0.011 ± 0.0002%	0.129 ± 0.003%

It is clear from the XRF of the feed samples shown in Figure 5.2 that the light elements (LE) form the bulk of the ore, with magnesium, aluminum, calcium, and chromium as other elements in significant quantities. The mean copper, nickel, and sulfur values shown in Table 5.1 for the selected UG2 ore were calculated from the concentrate and tailings data developed from all the batch flotation tests.

5.2.2 Plant Water Preparation

As aforementioned, the Centre for Minerals Research (CMR) at the University of Cape Town (UCT) developed a synthetic plant water (1 SPW) recipe to mimic typical water quality determined at several PGM concentrators across the BIC [51]. At the time, this water comprised of total dissolved solids content (TDS) of 1023 mg/L and ionic strength (IS) of 0.0242 mol.dm^{-3} [13, 51]. However steady increases in dissolved ion concentrations in process, recycled and tailings water has rendered water of a TDS of 1023 mg/L atypical of most PGM concentrators today [45]. Therefore, synthetic plant water of higher TDS content was proposed to better represent typical water concentrations in PGM concentrators.

Individual salts of analytical grade were dissolved in tap water to prepare synthetic plant water. The 3 SPW recipe was adapted from 1 SPW by multiplying the concentration of respective ions by three times to represent this increase in ion concentration on-site over time. All salts were supplied by Merck in powder form and were assumed to be of 99% purity. The salts were dissolved in tap water with an impeller in two 3-L buckets. Three SPW was considered to represent the current on-site feed process water quality, and was used as the baseline water quality for this study. Standards ionic concentrations contained in 1 and 3 SPW are shown in Table 5.2.

5.2.3 Reagents Preparation, Storage, and Disposal

A total of five reagents were utilized in all the experiments in this study. These are collectors, frothers, and depressants during the flotation procedure, while phenol and sulfuric acid were used during the colorimetric

Table 5.2 The concentration of ions (in mg/L) present in synthetic plant water (adapted from Wiese *et al.* [51], Corin *et al.* [13], and Manono *et al.* [45]).

	Ca^{2+}	Mg^{2+}	Na^{+}	Cl^{-}	SO_4^{2-}	NO_3^{-}	CO_3^{2-}	TDS (mg/L)	Ionic strength (M)
1 SPW	80	70	153	287	240	176	17	1023	0.0241
3 SPW	240	212	459	861	719	527	51	3069	0.0721

analysis for residual depressant in the tailings and concentrate water, following the du Bois *et al.* [55] method. The properties and suppliers of these reagents are outlined in Table 5.3.

5.2.3.1 Collector

The ionic thiol collector utilized in all the batch flotation tests was sodium isobutyl xanthate (SIBX), with a 90 % active content at a molecular weight of 172 g/mol supplied by AECI Senmin in pellet form. Due to the rapid decomposition of xanthates at elevated temperature, a 1% (w/v) solution of SIBX was prepared daily by crushing the pellets in a crucible and motor, and dissolving the solid in distilled water before experiments. The collector was dosed into the mill at 150 g/t (corrected for its active content) for all the runs. When the reagent was not in use, it was stored in a fridge at 9°C. The residual SIBX solution was discarded daily as outlined in the product material and data safety sheet and laboratory guidelines.

5.2.3.2 Depressant

The polysaccharide depressant utilized in the batch flotation tests was a sodium carboxy methyl cellulose (CMC) of a molecular weight of 325,000 g/mol, and was supplied at a purity of 72% by Akzo Nobel Functional Chemicals in powder form. A 1% (w/v) solution of the depressant was prepared daily for use by hydrating the required amount of dry depressant powder in distilled water. The contents of the solution were agitated using a magnetic stirrer for two hours to prevent the formation of lumps. The depressant was dosed at 0 g/t, 100 g/t, and 300 g/t, as prescribed in the experimental design. The depressant dosages of 300 g/t and 0 g/t were used to determine the entrainment factor based on the assumption that at an extremely high dosage, all-naturally floatable gangue (NFG) is depressed, and the only source of gangue is through entrainment in the froth. The 100 g/t CMC dosage was the normal dosage for all the recirculation

Table 5.3 Summary of the properties of flotation reagents utilized in the experiments.

Collector	**Type**	**SIBX**	
	Dosage	150	g/t
	Molecular weight	172	g/mol
	Chemical formula	$(CH_3)CHCH_2OCS_2Na$	
	Purity	90%	
	Supplier	AECI Senmin	
Frother	**Type**	**DOW 200**	
	Dosage	40	g/t
	Molecular weight	220	g/mol
	Chemical formula	$CH_3(C_3H_6O)_3O$	
	Purity	100%	
	Supplier	DOW Chemical Company	
Depressant	**Type**	**CMC (Depramin 267)**	
	Dosage	0, 100, 500	
	Molecular weight	325,000	g/mol
	Chemical formula	$C_8H_{15}NaO_8$	
	Purity	72%	
	Supplier	AKZO Nobel Functional Chemicals	

experiments. When the reagent was not in use, it was stored in a fridge at 9°C. The residual Depramin 267 solution was discarded as outlined by the product material data safety sheet and laboratory rules.

5.2.3.3 Frother

The polyglycol frother utilized in the batch flotation tests was DOW 200, supplied by the DOW Chemical Company, of a molecular weight of 220 g/mol with a purity of 100% in liquid form. The frother was dosed in the mill at 40 g/t for all runs using a P200 Gilson Air displacement pipette. After use, the reagent was stored at room temperature in the reagent cupboard.

5.2.4 Comminution of the UG2 Ore

The 1.3-kg ore samples prepared in Section 5.2.1 were ground in a laboratory-scale stainless steel rod mill (10-L Eriez MACSALAB Rod Vessel Mill of Grade 304 driven by a double roller of 120-mm diameter by 1200-mm length powered by a 0.37-kW 220-Volt Motor (Eriez Manufacturing Co., 2021)) to 60% passing 75 μm at 66% solid mass content by adding 681 ± 3 g of feed water (3 SPW or the recycled water relevant to the particular test). The Eriez stainless steel mill, milling media, and dimensions are shown in Table 5.4. The ionic thiol collector described above was added to the mill followed by the addition of 3 SPW plant water or recycled water as described above. Finally, the 1.3-kg ore samples were added to the mill followed by the addition of the rods as described in the mill specifications summary in Table 5.4. The mill lid was placed on and screwed tightly to prevent spillages. The mill was then placed on a double roller, which was turned on at 62 rpm, and a timer was started simultaneously. The mill was stopped after the milling time of 1317 s was reached, and the pulp was recovered to be transferred to the float cell.

The milling time, targeting 60% passing 75 μm was determined and found to be 21 min and 57 s (1317 s). The grind size was chosen as it

Table 5.4 Characteristics for the 10-L stainless steel laboratory scale mill.

Rods	**Diameter**	**Length**	**Number of rods**
	25 mm	285 mm	6
	20 mm	285 mm	8
	16 mm	285 mm	6
Mill	**Internal diameter**	**Internal length**	**Rotation speed**
	210 mm	295 mm	62 rpm
Calculated milling time	1317 s		

matched the rougher grind used in mining operations processing UG2 ores similar to the one used in this study.

5.2.5 Batch Flotation Procedure

The bench-scale flotation tests were conducted in a 3-L Barker laboratory flotation cell. The flotation standard conditions and variables of the cell are shown in Table 5.5. The milled slurry was transferred to the 3-L Barker flotation cell with a top-driven impeller, and a Wilkerson ¼ inch 0-8 bar air regulator to control the air flow rate. The impeller speed was set to 1200 rpm. The flotation cell containing the slurry was topped up to 3 L with the necessary water to achieve a pulp density of 33% solids. A 40-mL feed sample was taken before the addition of the flotation reagents to the float cell using a syringe. The slurry was conditioned for 2 min with the CMC depressant, Depramin 267, at the required dosage followed by 1 min with the polyglycol frother, DOW 200 at a dosage of 40 g/t.

For the experiments with varying depressant dosage (0 g/t CMC, 100 g/t CMC, and 300 g/t CMC), freshly prepared 3 SPW was used, whereas for those experiments with simulated short water recirculation (to the mill or to the float cell), CMC depressant was dosed at a fixed dosage of 100 g/t. After reagent conditioning, the air valve was opened and a constant airflow rate of 7 L/min was maintained for all tests. Four concentrates, denoted as

Table 5.5 Summary of the batch flotation procedure.

Action	Conditioning	Cumulative time (min)
Milling	As shown in Table 5.4	As shown in Table 5.4
Collector	Added to the mill	Added to the mill
Depressants	2	0
Frother	1	2
Air fed	0	3
C1	2	5
C2	4	9
C3	6	15
C4	8	23

C1 to C4, were collected through the scraping of the froth phase at 15-s intervals, for 2, 4, 6, and 8 min, respectively, while maintaining a 2-cm froth height throughout by the manual addition of plant water of 3 SPW quality. Two 40-mL tailings samples were taken after the flotation process using a syringe. The general experimental sequence is summarized in Table 5.5 and Figure 5.3.

The water recovered with the feed, concentrates, and tailings samples was measured and recorded. The feed sample, concentrates, and tailings samples were filtered using a Buchner funnel. The solids were dried in an oven overnight and weighed before performing chemical analysis, copper and nickel assays, through XRF analysis. Appreciating that this may not be perfectly accurate with regard to the mineralogy of the sample, the gangue mineral concentration was determined from the difference between the total valuable mineral and the total mass of the concentrates.

The residual water from the concentrate samples was recovered from the Buchner flask. The tailings were filtered using a filter press through a filter paper. The tailings water was recovered from the water disposal pipe of the filter press to be used in the recirculation experiments. A 40-mL

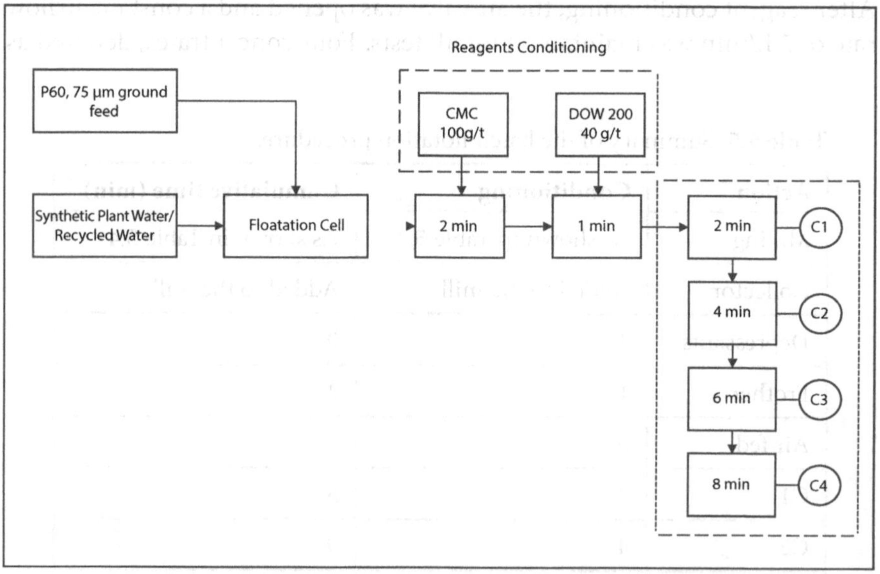

Figure 5.3 A summary of the experimental sequence adopted for the flotation of UG2 ore.

sample of the tailings and concentrate water was syringed and filtered through a 0.45-μm polyvinylidene fluoride syringe filter, and stored in a 50-mL sample vial. These samples were used for the chemical analyses to determine the ion concentration of selected ions present in the solution through GDACA.

It is worth mentioning that no reagents were utilized for pH maintenance, during the flotation test. However, the pH over all the experiments was at 8.23±0.03, typical of a low-grade Cu–Ni–PGM sulfide ore flotation [15].

5.2.6 Determination of the Entrainment Factor and Gangue Recovery by Entrainment

Entrainment is the non-selective recovery of both valuable and gangue particles carried up by the flow of water and air into the concentrate [56]. Water recovery has been associated with the degree of entrainment, otherwise known as the entrainment factor (ε), as water is the medium of transport and is related to the entrainment factor through Equation 5.1 where R_{gangue} and R_{water} represent the recovery of gangue and that of water, respectively. Thus, in order to determine the entrainment factor, Equation 5.1 was utilized based on the mass balance from Phase 1 of the flotation process experiments. The mass balance over Phase 1 tests allowed for the determination of the amount of water and gangue recovered at a CMC dosage of 300 g/t, using Equation 5.1, this further allowed for the prediction of ε. Using ε across all flotation experiments conducted in this study, the amount of entrained gangue recovered was determined.

$$R_{gangue} = \varepsilon R_{water} \tag{5.1}$$

5.2.7 Simulating Short Water Recirculation

When simulating short-term recycle runs, following initial flotation, the tailings filtrate was recovered using a filter press and used as part of the feed water for subsequent flotation. The total volume of tailings water recovered and used in subsequent flotation was 2200 ± 34 g. There were two routes in which the water was utilized. Mill recirculation as shown in Figure 5.4 involved the recirculation of 681 ± 3 g of the tailings water being fed to the mill to be used as the process water for milling. The rest of the recirculated water, 1519 ± 34 g, was utilized in washing the mill and transferring the

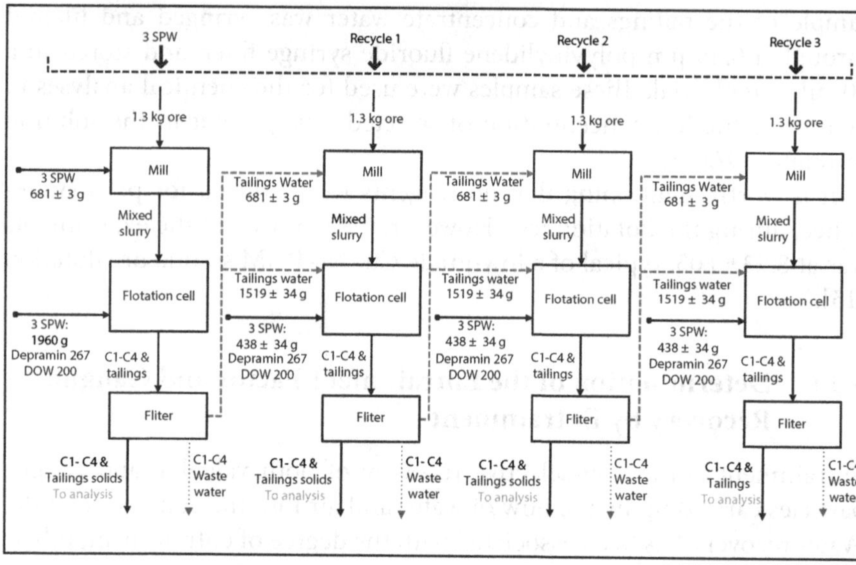

Figure 5.4 Simulation of short water recirculation to the mill.

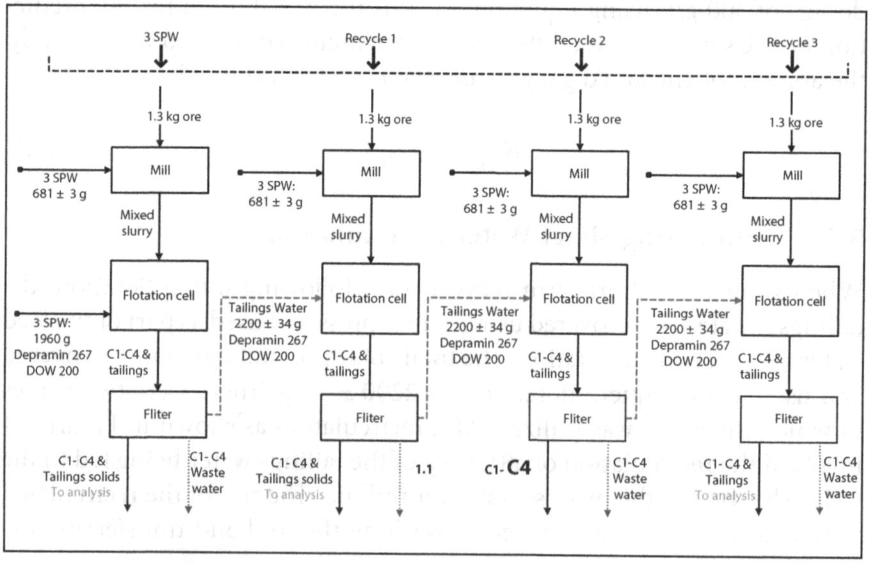

Figure 5.5 Simulation of short water recirculation float cell.

slurry to the float cell. An additional 438 ± 34 g of 3 SPW was used to top up the float cell to the 3-L mark (to maintain a 33 % pulp density).

The second route was float cell recirculation as shown in Figure 5.5 where 681 ± 3 g of 3 SPW water was fed to the mill to be used as process water for milling. Of the 2200 ± 34 g of recirculated water recovered, 1920 ± 34 g was then used to transfer the slurry to the float cell and the rest of the recirculated water, 280 ± 34 g, was used as top up water during the flotation experiment to its completion. The water recirculation procedure was repeated three times for both points of addition to create recycle 1, 2, and 3. It is worth noting that a similar approach to the procedures shown in Figure 5.4 and Figure 5.5 had been employed by Sinche-Gonzalez *et al.* [10, 11] on an investigation of pyrite and molybdenite flotation in process water containing organics.

5.2.8 XRF Analysis of Solids Samples

The metal ion content in the flotation concentrate, feed, and tailings was determined using an Olympus Vanta Handheld-XRF Analyzer. All solid samples were analyzed for their copper and nickel concentrations, which were utilized to develop the mass balance over the flotation process.

5.2.9 Thermo Scientific Gallery Discrete Automated Photometric (Colorimetric) Analyser (GDAPA)

The study also focused on the monitoring of the ionic species present in solution using a Gallery Discrete Automated Photometric (Colorimetric) Analyser (GDAPA). The ions chosen are of significant quantities in the ore body as indicated from the XRF analysis of the ore shown in Figure 5.2. Individual elemental composition of the UG2 ore feed was analyzed using an X-ray fluorescence (XRF) as aforementioned. GDAPAs have been used by water quality analysts and regulators to develop an understanding of complex water matrices. This technique is used, primarily, to monitor increased inputs of total dissolved solids (TDS) and specific ions in water resulting from mining activities, wastewater effluent discharge, agricultural practice, irrigation runoff, and rising/brackish groundwater, estuarine water, and nutrient cycling in hypersaline environments [57, 58]. GDAPAs are fast, flexible, and precise for complex samples of variable matrixes, test parameters, and low volume samples, achieving a high throughput with minimal supervision [59]. All analyses, for this study, were performed on a Thermo Fisher Scientific Automated Gallery Discrete Colorimetric Analyser. The instrument is calibrated online using stock calibrators stored on the analyzer as shown in Table 5.6 with reagent compositions that are all in compliance with the ISO 15923-1 standard.

Table 5.6 Analytes, reagents, and reference standards [59].

Product/analyte	Composition of system reagent	Reference standard
Chloride R1 (984264)	Methanol, iron (3) nitrate, mercury di-thiocyanate, nitric acid	ISO 15923-1 ISO 15682
Sulfate R1 (984648)	Barium chloride, hydrochloric acid, stabilizers	ISO 15923-1 SM4500 SO_4^{2-}E
Magnesium R1 (984358)	Xylidyl blue I	Tietz Fundamentals of Clinical Chemistry, 5th Edition
Calcium R1 (984361)	Arsenazo III	Tietz Fundamentals of Clinical Chemistry, 5th Edition

Stetson *et al.* [60] summarize the operations in the analyzer as follows; the incubator hub aligns the cuvette segment along the sample and reagent discharge paths. The cell segments are then aligned one after another with the dosing needle of the robot arm. The samples are then mixed using an ancillary stirrer and then returned to the washing station where the needle and stirrer blades are washed with deionized water. During the dosing process, the cuvette segment returns to the incubator, where a programmed reaction time of up to 3600 s occurs. Reagent blanks are measured after the sample and the first reagent are mixed. The incubation program finalizes in the photometric module. The modules sequentially measure the absorbance of chromophores in each cell of the cuvette segment. The absorbance of the reagent blanks is converted to concentrations from on-line calibration parameters using regression parameters. The parameters are sourced from the instruments operating system database memory.

5.3 Results and Discussion

5.3.1 Solids and Water Recoveries

Figure 5.6 shows the overall recovery of solids and process water reporting to the concentrates while Table 5.7 displays the amount of entrained

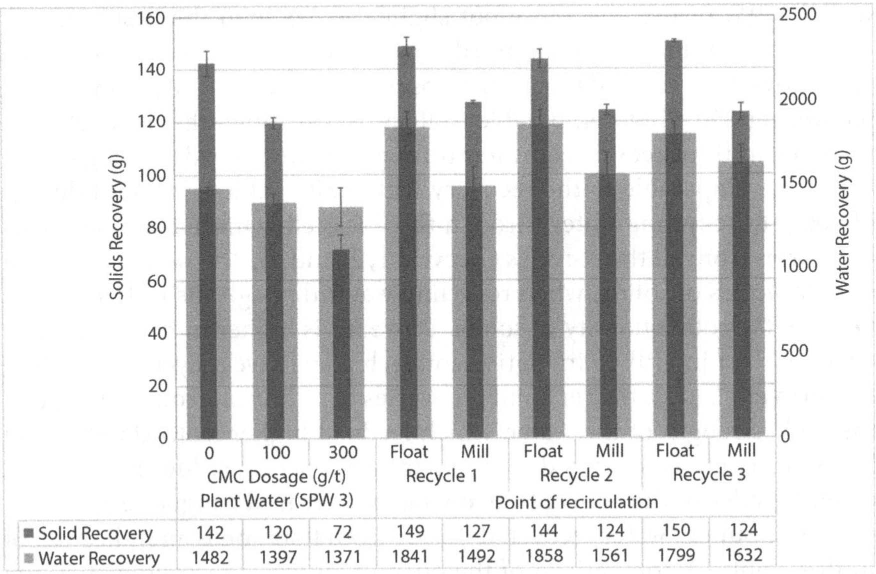

Figure 5.6 The quantity of solids and water reporting to the concentrate for all conditions tested.

gangue and ratio of solids to water recovery at all the different conditions tested in this study.

The first phase of the flotation experiments carried out in this study evaluates the impact of changing the depressant dosage on what. The quantity of solids recovered across the first phase descends from 142 g to 120 g to 71.6 g at depressant dosages of 0 g/t, 100 g/t, and 300 g/t, respectively, showing a decrease with increasing depressant dosage. The quantity water recovered across the first phase descends from 1480 g to 1400 g for floats at depressant dosages of 0 g/t and 100 g/t, respectively, but is constant between 100 g/t and 300 g/t.

Flotation at 100 g/t depressant, 40 g/t frother, and 150 g/t collector dosages was a benchmark to assess the deviation caused by recycling to flotation performance, and is the baseline (Recycle 0) for the recycles. Comparing the solids recovery over the first two recycling points, it can be seen that a higher solids recovery at the first float cell recycle, recovering 149 g, was achieved compared to the first mill recycle that recovered 127 g of solids. Overall, they are higher than the solid recovery observed at a 100 g/t depressant dosage of 120 g with no recycle (3 SPW). At the second recycle, the quantity of solids recovered in the second float recycle remained higher than that observed at 100 g/t depressant dosage with 3 SPW, recovering 144 g but slightly lower than the first float cell recycle by 5 g. Finally, at the third recycle, the third recycle to the float cell recovers more solids than that observed at 100 g/t CMC and 3 SPW, recovering 150 g but comparable to Recycle 1 through the float cell. The second and third recycles of water through the mill, yielded 124 g of solids each, comparable to the recovery with 3 SPW at a depressant dosage of 100 g/t. Recycling water through the float cell achieved similar solids recoveries from all the recycles (Recycle 1, 2, and 3). The same pattern on the recoveries of solids, when recycling water through the mill, is noted. Overall, the total quantity of solids recovered is higher in float cell recirculation than in mill recirculation for each respective recycle. According to Corin *et al.* [13], higher solid recoveries could result from a change in the ionic strength of the water, which leads to higher froth stability and higher recoveries. Increased solids recoveries may cause lower nickel and copper grades in the concentrate owing to increased gangue recoveries.

The amount of water recovered during froth flotation is an indication of froth stability. A comparison of the quantity of water recovered at all the recycles against each other and against the water recovery achieved when floating using 3 SPW at a depressant dosage of 100 g/t was thus conducted. Analyzing the water recovery when floating with once recycled water, it can be noted that the quantity of water recovered when recycling to the float cell is higher, with a water recovery of 1840 g/t than that observed with 3 SPW at 100 g/t depressant dosage, recovering 1400 g.

For the second recycle, we note that the quantity of water recovered when recycling through the float cell is also higher than that observed when floating with 3 SPW at 100 g/t depressant dosage, recovering 1858 g of water. Finally, in the third recycle, the water when floating with recycled water, at both recycling points, is more than that recovered when floating using 3 SPW 100 g/t depressant dosage, recovering 1800 g and 1630 g when recycling through the float cell and mill, respectively. Overall, the quantity of water recovered is higher when recycled water is introduced into the flotation cell than floating with 3 SPW at the same fresh feed-based depressant, frother, and collector dosages.

Recycling through the mill also achieved higher water recoveries compared to floating with 3 SPW at the same depressant, frother, and collector dosage feed. The increase in water recoveries was in the order Recycle 3>Recycle 2>Recycle 1>3 SPW, as was the case with float cell recirculation. However, water recycling to the float cell resulted in higher water recoveries compared to water recycling to the mill. The higher water recoveries in float cell recirculation indicate a higher froth stabilization compared to mill recirculation. The increase in froth stability of the float cell recycle water could be because of an increase in the ionic concentration, and the presence of residual frother and it being more active when recycled water is added to the float cell. Manono *et al.* [61] observed that increasing the ionic strength of water, and hence the concentration of ions, increased the water recoveries during flotation.

Table 5.7 shows the amount of entrained gangue and solids to water ratio for all the flotation experiments carried out. For the experiments conducted using 3 SPW, the amount of entrained gangue decreased with an increase in the depressant dosage. This result can be attributed to the reduction in the amount of water recovered, and therefore froth stability [13, 45, 62]. The decrease in solids-to-water ratio suggests that less solids were pulled to the concentrate per gram of water in increasing depressant dosage. This result has been seen by Corin *et al.* [13]. The decrease in water recovery accompanied by the reduction in entrained gangue from 77.4 g to 72.9 g between floats at 0 and 100 g/t depressant dosages aligns well with the observed decline in solids recoveries. The decrease in solids recoveries observed between floats at 100 and 300 g/t also ties nicely with the reduction in entrainment from 72.9 g to 71.6 g.

Table 5.7 also shows that the amount of engrained gangue was higher in float cell recirculation compared to mill recirculation. However, overall recirculated process water resulted in higher amounts of entrained gangue compared to 3 SPW. This trend is in line with the higher solids and water recoveries, shown in Figure 5.5, in recirculated water compared to the use of 3 SPW. It is also interesting to note that the solids-to-water ratio did not change with the number of recycles.

Based on Table 5.7, overall, increasing the depressant dosage led to an overall decrease in solids and the entrainment gangue, showing the selective impact of the depressant on gangue and improving their hydrophilicity, led

Table 5.7 Entrained gangue (Ent. gangue) and solids to water ratio for all the conditions tested.

	Plant water (3 SPW)			**Point of recirculation**					
	CMC dosage (g/t)			**Recycle 1**		**Recycle 2**		**Recycle 3**	
	0	**100**	**300**	**Float**	**Mill**	**Float**	**Mill**	**Float**	**Mill**
Ent. gangue (g)	77.4	72.9	71.6	96.1	77.9	97	81.5	92.2	85.2
S/W (%)	10	9	5	8	8	9	8	8	8

to lower solids recoveries. This results in the expectation of an increased grade and recovery of copper and nickel in the concentrates, with more of the gangue reporting to the tailings. A destabilization of the froth, resulting from the addition of the depressant, leads to better drainage in the froth phase and coagulation of gangue into the pulp phase, improving the selectivity of the froth flotation process as shown by Corin *et al.* [13] and Wiese *et al.* [63], who had similar trends in grades and recoveries with increased depressant dosage, as observed in this study. Although the entrained gangue increases in the order Float Cell Recycles<Mill Recycles<3 SPW, and the solids-to-water ratio achieved in float cell and mill recycles is similar to that obtained with 3 SPW (100 g/t CMC), implying that the depressant acts primarily on the mineral surface, inducing hydrophilicity.

The mass of solids recovered across all the tests, conducted at 100 g/t depressant, 40 g/t frother, and 150 g/t collector dosage at varying water quality as a function of time, are shown in Figure 5.7. First order rate models were visually fit to the data, developing four distinct rate relationships between floating time and solids recovery. R_{max} represents the maximum possible solids recovery while k represents the rate constant (in grams per second of solids recovered), hence a higher rate constant means the process approaches its maximum quicker, hence a lower residence time would be needed for a required mass pull. The highest rate of solids recovery is observed during the final recycle to the float, and has the highest possible solids recovery and the largest k value of $0.169*10^{-2}$ s^{-1}. The second-rate

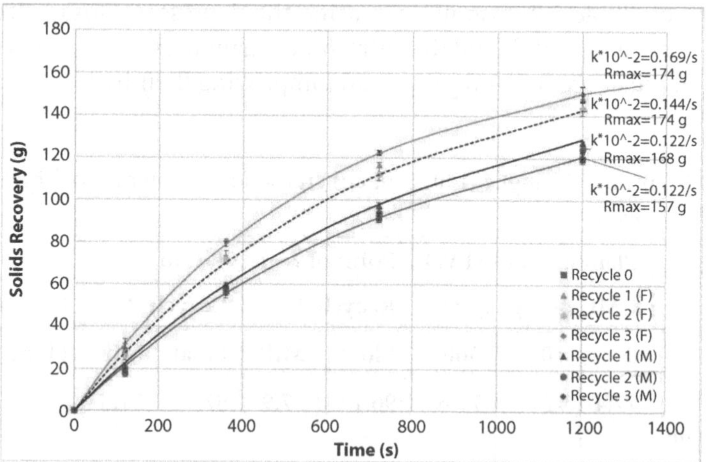

Figure 5.7 Cumulative solids recovery across all the floats from the second phase of experiments as a function of time.

model has a lower k value of $0.144*10^{-2}$ s^{-1}, and well fits Recycle 1 and 2 through the float cells. This is as expected, as recycling through the float recovered more solids overall. The third model achieves a total solid recovery of 168 g, at a k value of $0.122*10^{-2}$ s^{-1}. Visually, this model best described mill recirculation tests, which achieved overall lower solids recoveries when compared to float recirculation. The last model has the lowest maximum solids recovery of 157 g, and visually models flotations using 3 SPW at the standard reagent dosages. From the rate models, water recirculation leads to a change in the flotation conditions, leading to an increase in solids recovery, more increased when recycling through the float cell than the mill. This increase in water recoveries and, therefore, greater stabilization of the froth may lead to increases in solids recoveries. This may, however, promote non-selective modes of recovery such as entrainment, and hence could negatively impact the grade of the concentrate.

5.3.2 Copper and Nickel Recoveries and Grades

Figure 5.8 shows the cumulative copper recovery and grade for all tested conditions.

Across the first phase of experiments, increasing concentration of the depressant between 100 and 300 g/t led to a significant increase in the grade by about three-fold from 0.062% to 0.164%. The similar recovery of

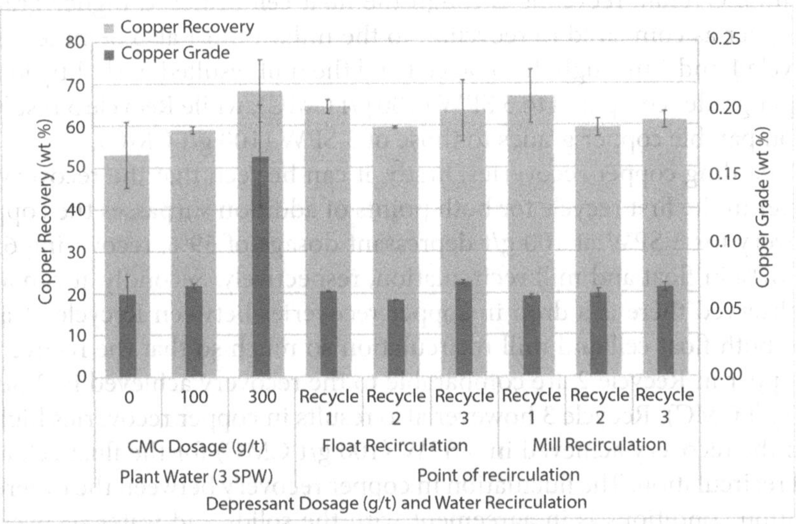

Figure 5.8 The recovery and grade of copper in the solids reporting to the concentrates for all tested conditions.

copper between 100 and 300 g/t dosage and the increase in copper grade indicates that the depressant enhances the selectivity of the froth flotation process through increasing the hydrophilicity of gangue minerals. Floating at 0 g/t achieved comparable copper recoveries to flotation with 100 and 300 g/t. We also observed a slight increase in recovery when increasing the depressant from 100 to 300 g/t from 60% to 68%, respectively. Thus, it can be affirmed that at a constant collector and frother dosage, an increase in the depressant dosage led to an increase in the grade of copper as compared to non-depressed flotation with a minimal impact on the recovery, showing that the interaction between depressant and gangue was enhanced in increased depressant dosages.

In the second phase of experiments, the water quality was varied through the use of recycled water consecutively recycled through either the mill or the float cell for three recycles. Copper grades and recoveries were compared against individual recycles and against flotation with 3SPW at 100 g/t depressant dosage, 40 g/t frother, and 150 g/t collector dosage (Recycle 0). At the first recycle stage, Recycle 1, recycling through the float cell achieved a slightly higher copper grade of 0.064%, as compared to recycling through the mill with a grade of 0.060%. Conversely, at the second recycle, recycling through the mill achieved a slightly higher recovery of 0.062% as compared to 0.057% when recycling through the float cell. In the third recycle, recycling through the float achieved a slightly higher grade of 0.071% as compared to 0.067% when recycling through the mill. Overall, recycling through the float cell achieved higher grades of copper as compared to recycling to the mill. It must also be noted that Recycle 1 and 2 through the float cell and the mill resulted in slightly lower copper grades compared to 3 SPW (100 g/t CMC), while Recycle 3 resulted in comparable copper grades to those of 3 SPW (100 g/t CMC).

Regarding copper recoveries, firstly, it can be seen that the recovery of copper in the first recycle for both points of addition surpasses the copper recovery for 3 SPW at 100 g/t depressant dosage of 59%, recovering 65% and 67% in float and mill recirculation, respectively. Secondly, it can also be observed there is a drop in copper recoveries between Recycles 1 and 2 for both float cell and mill recirculation so much so that the recoveries of copper in Recycle 2 are comparable to the recovery achieved in 3 SPW (100 g/t CMC). Recycle 3 however also results in copper recoveries higher than the recovery achieved in 3 SPW (100 g/t CMC) for the float cell and mill recirculation. The fluctuation in copper recovery between the different flotation conditions is in agreement with the solids and water recoveries achieved, and can thus be attributed to changes in froth stability.

Figure 5.9 shows the cumulative final nickel recovery and grade for all tested conditions.

At a constant collector dosage of 150 g/t and a frother dosage of 40 g/t, increasing the depressant dosage from 0, 100, and 300 g/t led to an increase in nickel grade from 0.218%< 0.246%< 0.412%, doubling between 100 and 300 g/t while nickel recoveries between these three runs were 18.2%, 19.2%, and 17.4% for 0, 100, and 300 g/t, respectively. The increase in nickel grades is in agreement with the trend of a decrease in solids and water recoveries, which is believed to be resulting from a decrease in froth stability, and thus a decrease in gangue recoveries with increasing depressant dosage. The decrease in nickel recovery at 3 SPW 300 g/t CMC can be attributed to the presence of pentlandite-gangue composites, i.e., pentlandite particles being depressed along with gangue particles due to their association [13].

The effect of recycling water at the first recycle leads to a higher recovery when recycling to the float cell at 20.8% compared to the use of 3 SPW (100 g/t CMC) while the first recycle to the mill resulted in similar nickel recoveries to those achieved with 3 SPW (100 g/t CMC). For the second recycle, the recovery of nickel with float cell water recycle is higher compared to freshly prepared synthetic plant water (3 SPW 100 g/t CMC). However, water recirculation to the mill resulted in nickel recoveries even lower than those achieved with 3 SPW 100 g/t. Finally, in the third recirculation, again the recovery of nickel in float cell recycle is higher than in

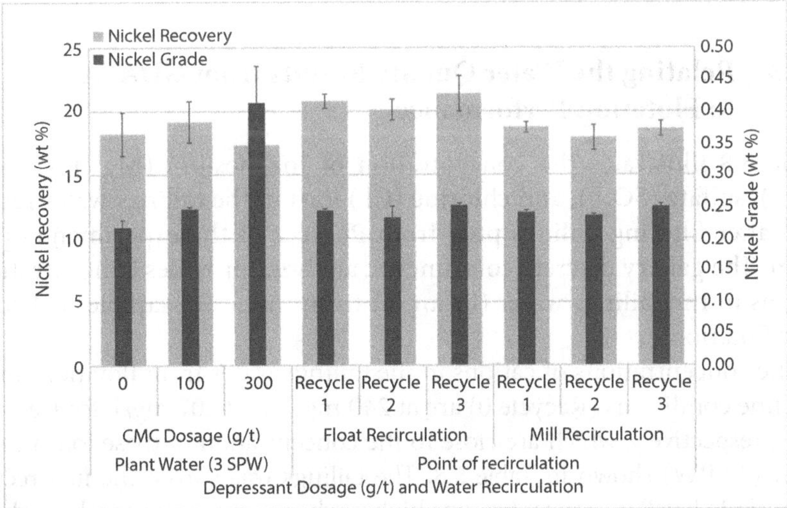

Figure 5.9 The recovery and grade of nickel in the solids reporting to the concentrates for all conditions.

mill recirculation and higher than that obtained with 3 SPW 100 g/t CMC; the order is as follows, Float Cell Recycle 3> Mill Recycle 3 = 3 SPW 100 g/t. A bank of concentrators recycling water to the float cell would likely result in higher nickel recovery compared to recycling water to the mill. On the other hand, recycling water to the mill is likely going to result in similar nickel recoveries as those obtained with 3 SPW, thus suggesting the importance of the point of addition for the flotation of pentlandite.

Conversely, the point of addition did not lead to any meaningful change in the grade achieved at similar recycle stages as it remains stable at 0.243 ± 0.080 % as with recoveries when floating with 3 SPW at similar reagent dosages. In fact, the only measurable change in nickel grade is between Recycle 1 and Recycle 3 where a higher grade is observed at Recycle 3 at 0.252 ± 0.003 % > 0.244 ± 0.002 % in float recirculation akin to 0.251 ± 0.004 % > 0.240 ± 0.003 % in mill recirculation.

It can be said that an increase in depressant dosage at a constant collector and frother dosage increased the nickel grade. At a depressant dosage of 100 g/t, water recirculation leads to a decrease in the overall recovery of nickel for mill recirculation, while an increase is seen for float cell recirculation and a slight decrease in the grade of nickel in the second recirculation, for both recirculation configurations, is seen.

Overall, the grades and recoveries of copper and nickel are low, and this has been attributed to the reagent suite chosen, which was not optimized for the ore, and the operating pH of 8.23 ± 0.03, which is attributed to lower recoveries and grades in the flotation of UG2 ores [64].

5.3.3 Relating the Water Quality Results from GDAPA to Flotation Performance

Table 5.8 illustrates the concentration of magnesium (Mg^{2+}), calcium (Ca^{2+}), sulfate (SO_4^{2-}), and chloride (Cl^-) ions in the tailings water recovered after filtering tailings pulp from Phase 2 of the experimental program. The gallery discrete colorimetric analyzer provides concentrations of ions in the tailings water (in mg/L) from the water samples collected after flotation.

The concentrations of cations in the tailings water from flotation under baseline conditions (Recycle 0) are at 240 mg/L and 207 mg/L for Ca^{2+} and Mg^{2+}, respectively, which are close to the concentration of these ions in feed water (3 SPW) shown in Table 5.8. The tailings water from the first recycle (Recycle 1) has Ca^{2+} concentrations higher when recycling water through the float cell at 248 mg/L, than when recycling water through the mill at 237 mg/L.

The tailings water from the second recycles (Recycle 2) has Ca^{2+} concentrations higher when recycling water through the float cell at 254 mg/L, than when recycling water through the mill at 234 mg/L. Finally, the tailings water from the third recycle (Recycle 3) has higher Ca^{2+} concentrations when water is recycled through the float cell at 258 mg/L, than when recycling water through the mill at 241 mg/L. A consistent trend for Ca^{2+} is that for a particular recirculation, the concentration is similar in all the tailings samples. Tailings water samples were also compared based on their magnesium concentrations. Tailings water from the first recycles (Recycle 1) has Mg^{2+} concentrations that are higher when recycling through the float cell at 242 mg/L compared to when recycling through the mill at 208 mg/L. The tailings water from the second recycle (Recycle 2) has Mg^{2+} concentrations higher when recycling through the float cell at 198 mg/L, than when recycling through the mill at 193 mg/L. Finally, the tailings water from the third recycles resulted in Mg^{2+} concentrations that are higher when recycling through the float cell at 191 mg/L, than when recycling through the mill at 164 mg/L. Overall, the concentration of Ca^{2+} and Mg^{2+} in tailings water is higher in float recirculation than in mill recirculation. Also of note, all the Ca^{2+} concentrations in mill recirculation are close to the ion's concentrations in the feed water (3 SPW). Mg^{2+} in the tailings in mill recirculation, and float cell recirculation, with exemption of Recycle 1 through the float cell, are lower than Mg^{2+} concentration in the feed water (3 SPW).

Calcium ions have been shown to adsorb on mineral surfaces by Ikumapayi and Rao [43], and lead to higher nickel recoveries, explaining the higher nickel recoveries observed in float cell recirculation than mill recirculation [65]. The increase in copper grades at higher Ca^{2+} concentrations was also observed by Dzingai *et al.* [15]. They showed an increase in copper and nickel recoveries, and grades at higher concentrations of Ca^{2+} spiked to 400 mg/L in 3 SPW as background matrix. Manono *et al.* [66, 67] also showed that increases in the ionic strength of plant water containing Ca^{2+} and Mg^{2+} result in an increase in the adsorption of CMC onto naturally floatable gangue, thus impeding the floatability of gangue and increasing the grades of the concentrate. This finding is also in line with work by Laskowski *et al.* [68].

Table 5.8 also shows the concentrations of anions in the tailings water, recovered after filtering tailings pulp from Phase 2 of the experimental program. The anions concentrations in the tailings water from flotation under baseline conditions (Recycle 0) are at 792 mg/L and 1250 mg/L, for SO_4^{2-} and Cl^- respectively, which are slightly higher than the concentration

Table 5.8 Concentration of dissolved Ca^{2+}, Mg^{2+}, SO_4^{2-}, and Cl^- present in tailings water recovered post-flotation in phase 2 of the experimental program.

			Recycle 1		Recycle 2		Recycle 3	
	3 SPW	Recycle 0	Mill	Float	Mill	Float	Mill	Float
Ca^{2+}	240	240 ± 3	237 ± 6	248 ± 1	234 ± 6	254 ± 2	241 ± 6	258 ± 6
Mg^{2+}	212	207 ± 3	207 ± 12	242 ± 7	193 ± 1	198 ± 3	164 ± 12	191 ± 12
SO_4^{2-}	719	792 ± 10	708 ± 14	804 ± 57	740 ± 9	805 ± 17	783 ± 14	813 ± 14
Cl^-	861	1247 ± 127	1057 ± 15	999 ± 120	1032 ± 12	1027 ± 128	1054 ± 15	1021 ± 15

of these ions in 3 SPW as shown in Table 5.8. The tailings water from the first recirculation has SO_4^{2-} concentrations higher when water is recycled through the float, at 804 mg/L, compared to recycling water through the mill, at 708 mg/L. The tailings water from the second recycle (Recycle 2) has SO_4^{2-} concentrations higher when recycling water through the float cell, at 805 mg/L, than when recycling water through the mill, at 740 mg/L. Finally, the tailings water from the third recycle has SO_4^{2-} concentrations higher in float cell recirculation, at 813 mg/L, than in mill recirculation, at 783 mg/L. Overall, the concentration of SO_4^{2-} in tailings water is higher in float cell recirculation than in mill recirculation; however, SO_4^{2-} tends to increase rapidly across the recycles in mill recirculation, while they remain similar in all float cell recirculation.

The higher solids recoveries in float cell recirculation can be linked to the higher sulfate concentrations that lead to more stable froths, which achieve higher mass pulls, explaining the disparity in recovery between float cell and mill recirculation [10, 11]. There is no association between sulfate concentrations and changes in the copper and nickel grades and recoveries; they fell within concentration boundaries of between 720 and 1200 mg/L previously attributed to being threshold concentrations that have no negative impact on the grades and recoveries [15].

Conversely, for Cl^- concentrations in tailings water, there was no discernible difference between similar recycle runs irrespective of the point of addition and remained stable in all the recycle stages. However, all Cl^- concentrations in the experiments with recycled water are lower than the measured chloride concentrations in the tailings water from the baseline floats, but higher than the calculated Cl^- in the 3 SPW recipe.

Overall, as float cell recirculated recycled' water has higher Ca^{2+}, Mg^{2+}, and SO_4^{2-}, it can be considered to have higher ionic strength, and hence leading to higher solids and water recovery due to an increase in froth stability similar to the observations of Dzingai *et al.* [15] and Dzingai *et al.* [69] who saw an increase in solids and water recoveries when moving from 3 SPW to 10 SPW and when spiking 3 SPW with Ca^{2+}, Mg^{2+}, and SO_4^{2-}.

5.4 Conclusions

Increased depressant concentrations showed a meaningful change in the flotation performance. It led to a decrease in solids recovery, with a slight decrease in water recovery. The decrease in solids recovery is because of the adsorption of depressant on gangue, improving and enhancing the hydrophilic nature and reducing the transport of gangue to the froth phase. This also leads to reduction of froth stabilizing gangue reporting to the froth phase, and hence, a less stable froth achieving a better drainage. As expected, similar improvements in performance are also observed in the copper and nickel grades by three-fold and two-fold, respectively, between the 100 and 300 g/t CMC dosages.

Water recirculation leads to a change in the water chemistry, and this inadvertently affected the flotation performance. The increase in ionic concentration of the water led to a change in the froth stability, leading to higher solids and water recoveries.

The point of addition led to a disparity in the observed mineral recoveries and grades. The higher solids recoveries in recirculation to the float cell correspond to higher overall electrical conductivity in the tailings water, promoting that recirculation to the float cell led to a higher change in the water quality. The point of addition leads to a clear disparity in the solids, water recoveries, and the grade and recovery of sulfides.

Float cell recirculation achieved more solids recovery compared to mill recirculation. Similarly, it also achieved a higher water recovery. Although the quantity of gangue reporting to float cell recirculation was expected to increase, this was not observed in the grades and recoveries of the copper and nickel in the concentrates. This could be because of the slow buildup of ions in a short-recirculated system, boosting stable water chemistry, more on the mill recirculation, whose performance in all factors remained relatively like the operation at 3 SPW conditions.

The higher overall grade of copper and recoveries of nickel observed in float cell recirculation could be because of the higher concentrations of Ca^{2+} in float recirculation. This improvement in performance could be because of the impact of magnesium and calcium adsorbing on the surface of talcaceous surfaces, reducing their hydrophobicity as observed in studies on the impact of ionic strength [13].

Overall, float cell recirculation leads to waters with higher concentrations of Ca^{2+}, Mg^{2+}, and SO_4^{2-} and are thought to be of high ionic strength, leading to the higher solids and water recoveries observed in float cell recirculation.

The water recirculation simulation allowed for an analysis of the ion's accumulation in relation to the electrical conductivity of the water. This informs on the impact of water recirculation on ore dissolution. The effect of water recirculation on the flotation performance and ore floatability allows prediction of the direction of the change. However, it is unable to forecast the grade or the recovery for real plant operations.

It is important to note that the water recirculation simulation may not display true plant dynamics as other factors such as process water quality onsite, ore variability, tailings ponds, makeup water quality, water blending strategies, operating conditions, and bacterial activities may have important effects on plant dynamics, process water quality on site, ore variability, tailing ponds, make up water quality, water blending strategy, operating conditions, and bacterial activities [50].

The water recirculation simulation described in this study could be an important tool in the prediction of the impact of water quality, and ore dissolution in plant operation with short, closed water circulation. Reproducibility of the water recirculation simulation results motivates its use as a complementary tool in traditional processing plant design and water treatment facility dimensioning for plants.

Acknowledgements

The authors would like to acknowledge members of the Reagent Research Group (RRG), particularly Impala Platinum for providing the ore sample. This research received funding from the UCT research and innovation office under the Building Research Active Academic Staff Grant (2020-2021). Other sources of funding that this research benefited from include the National Research Foundation (Grant no. 118062– Project Title: Water Quality in Flotation: Causes, Impacts and Process Management Strategies; Grant no. 138294 – Project Title: Investigating the Effects of Flotation Reagents on Tailings Dewatering).

References

1. Statistics South Africa, *Electricity, Gas and Water Supply Industry*, 2013.

2. Duncan, J., *CDP Water Disclosure 2010 South Africa Focus*, Carbon Disclosure Project, 2010.
3. Biçak, Ö., Ekmekçi, Z., Can, M., Öztürk, Y., The effect of water chemistry on froth stability and surface chemistry of the flotation of a Cu-Zn sulfide ore. *Int. J. Miner. Process.*, 102–103, 32–37, 2012.
4. Ikumapayi, F., Makitalo, M., Johansson, B., Rao, K.H., Recycling of process water in sulphide flotation: Effect of calcium and sulphate ions on flotation of galena. *Miner. Eng.*, 39, 77–88, 2012.
5. Boujounoui, K., Abidi, A., Bacaoui, A., El Amari, K., Yaacoubi, A., The influence of water quality on the flotation performance of complex sulphide ores: Case study at Hajar Mine, Morocco. 1243–1251, 2015.
6. Boujounoui, K., AbidiI, A., Bacaoui, A., El AmariI, K., Yaacoubi, A., The influence of water quality on the flotation performance of complex sulphide ore: Case study at Hajar Mine, Morocco. *J. South Afr. Inst. Min. Metall.*, 115, 1243–1251, 2015.
7. Forssberg, K.S.E. and Hallin, M.I., Process water reticulation in a lead-zinc plant and other sulphide flotation plants, in: *Proceedings of the In Proc, Symp. Challenges in Mineral Processing*, Society of Mining Engineers, 1988.
8. Rao, S.R. and Finch, J.A., A review of water re-use in flotation. *Miner. Eng.*, 2, 65–85, 1989.
9. Schumann, R., Levay, G., Dunne, R., Hart, S., *Managing Process Water Quality in Base Metal Sulfide Flotation*, pp. 13–15, 2003.
10. Sinche-Gonzalez, M. and Fornasiero, D., Understanding the effect of sulphate in mining-process water on sulphide flotation. *Miner. Eng.*, 165, 106865, 2021.
11. Sinche-Gonzalez, M., Fornasiero, D., Zanin, M., Flotation of chalcopyrite and molybdenite in the presence of organics in water. *Minerals*, 6, 1–13, 2016.
12. October, L., Corin, K., Schreithofer, N., Manono, M., Wiese, J., Water quality effects on bubble-particle attachment of pyrrhotite. *Miner. Eng.*, 131, 2019.
13. Corin, K.C., Reddy, A., Miyen, L., Wiese, J.G., Harris, P.J., The effect of ionic strength of plant water on valuable mineral and gangue recovery in a platinum bearing ore from the merensky reef. *Miner. Eng.*, 24, 131–137, 2011.
14. Manono, M., Corin, K., Wiese, J., Perspectives from literature on the influence of inorganic electrolytes present in plant water on flotation performance. 54, 1191–1214, 2018.
15. Dzingai, M., Manono, M., Corin, K., Simulating the effect of water recirculation on flotation through ion-spiking: effect of Ca2+ and Mg2+. *Minerals*, 10, 1033, 2020.
16. Slatter, K.A., Plint, N.D., Cole, M., Dilsook, V., De Vaux, D., Palm, N., Oostendorp, B., Water management in anglo platinum process operations: Effects of water quality on process operations. *International Mine Water Conference*, Pretoria, South Africa, October 19th–23rd 2009 Proceedings, pp. 46–55, 2009.

17. Levay, G., Smart, R., Skinner, W., The impact of water quality on flotation performance. *J. South Afr. Inst. Min. Metall.*, 1, 69–76, 2001.
18. Muzenda, E., An investigation into the effect of water quality on flotation performance. *World Acad. Science, Eng. Technol. Int. J. Chemical, Molecular, Nuclear, Mater. Metallurgical Eng.*, 4, 562–566, 2010.
19. Muzinda, I. and Schreithofer, N., Water quality effects on flotation: Impacts and control of residual xanthates. *Miner. Eng.*, 125, 34–41, 2018.
20. Manenzhe, R.M., Corin, K.C., Wiese, J.G., Manono, M.S., Musuku, B., The effect of water quality on the adsorption of a xanthate collector in the flotation of a sulphide ore, in: *Proceedings of the IMPC 2018-29th International Mineral Processing Congress*, 2019.
21. Musuku, B. and Dahl, O., Impact of recycled process water on electrochemical reactivity of sulphide ore. *Minerals*, 12, 1455, 2022.
22. Finkelstein, N.P. and Lovell, V.M., Fundamental studies of the flotation process: The work of the national institute for metallurgy. *J. South Afr. Inst. Min. Metall.*, 15, 1972.
23. Ikumapayi, F., Johansson, B., Rao, K.H., Recycling of process water in sulfides processing and flotation selectivity. *XXV International Mineral Processing Congress*, pp. 4079–4088, 2010.
24. October, L.L., *Water Quality Effects on the Bubble-Particle Attachment of Sulfide Minerals*, A Thesis Submitted to the Faculty of Engineering and the Built Environment, University of Cape Town, 2020.
25. Michaux, B., Rudolph, M., Reuter, M.A., Challenges in predicting the role of water chemistry in flotation through simulation with an emphasis on the influence of electrolytes. *Miner. Eng.*, 125, 252–264, 2018.
26. Chimonyo, W., Wiese, J., Corin, K., O'Connor, C., The use of oxidising agents for control of electrochemical potential in flotation. *Miner. Eng.*, 109, 135–143, 2017.
27. Chimonyo, W., Corin, K.C., Wiese, J.G., O'Connor, C.T., Redox potential control during flotation of a sulphide mineral ore. *Miner. Eng.*, 110, 57–64, 2017.
28. Mailula, T.D., Bradshaw, D.J., Harris, P.J., The effect of copper sulphate addition on the recovery of chromite in the flotation of UG2 ore. *J. South Afr. Inst. Min. Metall.*, 143–146, 2003.
29. Martinovic, J., Bradshaw, D.J., Harris, P.J., Investigation of surface properties of gangue minerals in platinum bearing ores. *J. South Afr. Inst. Min. Metall.*, 105, 349–356, 2005.
30. Khraisheh, M., Holland, C., Creany, C., Harris, P., Parolis, L., Effect of molecular weight and concentration on the adsorption of CMC onto talc at different ionic strengths. *Int. J. Miner. Process.*, 75, 197–206, 2005.
31. Finkelstein, N.P., Addendum to: The activation of sulphide minerals for flotation: A review. *Int. J. Miner. Process.*, 55, 283–286, 1999.

32. Haran, N.P., Boyapati, E.R., Boontanjai, C., Swaminathan, C., Kinetics studies on effect of recycled water on flotation of copper tailings from Benambra Mines, Victoria. *Dev. Chem. Eng. Miner. Process.*, 4, 197–211, 2008.
33. Wang, B. and Peng, Y., The effect of saline water on mineral flotation–a critical review. *Miner. Eng.*, 13–24, 2014.
34. Peters, E.N. and Meybeck, M., Water quality degradation effects on freshwater availability: Impacts of human activities. *Water Int.*, 25, 185–193, 2000.
35. Carlson, L., Bigham, J.M., Schwertmann, U., Kyek, A., Wagner, F., Scavenging of as from acid mine drainage by schwertmannite and ferrihydrite: A comparison with synthetic analogs. *Environ. Sci. Technol.*, 36, 1712–1719, 2002.
36. Johnson, N.W., *Issues in Maximisation of Recycling of Water in a Mineral Processing Plant*, The Australian Institute of Mining and Metallurgy, pp. 239–245, 2003.
37. Peng, Y., Jin, D., Li, J., Wang, C., Flocculation of mineral processing wastewater with polyacrylamide. *IOP Conf. Ser. Earth Environ. Sci.*, 565, 2020.
38. Corin, K.C. and Wiese, J.G., Investigating froth stability: A comparative study of ionic strength and frother dosage. *Miner. Eng.*, 66, 130–134, 2014.
39. Bruckard, W.J., Sparrow, G.J., Woodcock, J.T., A review of the effects of the grinding environment on the flotation of copper sulphides. *Int. J. Miner. Process.*, 100, 1–13, 2011.
40. Ma, X., Bruckard, W.J., Holmes, R., Effect of collector, PH and ionic strength on the cationic flotation of kaolinite. *Int. J. Miner. Process.*, 93, 54–58, 2009.
41. Ikumapayi, F., Recycling process water in sulfide flotation, Part B: Effect of H2O2 and process water components on sphalerite flotation from complex sulfide. 29, 192–198, 2013.
42. Ikumapayi, F., Sis, H., Johansson, B., K. H.Rao, Recycling process water in sulfide flotation, Part B: Effect of H2O2 and process water components on sphalerite flotation from complex sulfide. *Miner. Metall. Process.*, 29, 192–198, 2012.
43. Ikumapayi, F. and Rao, K.H., Recycling process water in complex sulfide ore flotation: Effect of calcium and sulfate on sulfide minerals recovery. *Miner. Process. Extr. Metall. Rev.*, 36, 45–64, 2015.
44. Ikumapayi, F.K., Johansson, B., Rao, K.H., Recycling of process water: Effect of calcium and sulphate ions in sulphides flotation. *Conference in Minerals Engineering: Luleå*, Februari 2-3, 2010, pp. 51–61, 2010.
45. Manono, M., Corin, K., Wiese, J., Water quality effects on a sulfidic pgm ore : Implications for froth stability and gangue management. 54, 1253–1265, 2018.
46. Abbassi, R., Khan, F., Hawboldt, K., Prediction of minerals producing acid mine drainage using a computer-assisted thermodynamic chemical equilibrium model. *Mine Water Environ.*, 28, 74–78, 2009.
47. Declercq, J., Charles, J., Bowell, R., Warrender, R., Barnes, A., *Comparison of Thermodynamic Equilibrium and Kinetic Approach in the Predictive Evaluation of Waste Rock Seepage Quality in Northern Finland.*

48. Muniruzzaman, M., Karlsson, T., Kauppila, P.M., Modelling tools for the prediction of drainage quality from mine wastes. *Bull. Geological Survey Finland*, 408, 27–42, 2018.
49. Le, T.M.K., Miettinen, H., Bomberg, M., Schreithofer, N., Dahl, O., Challenges in the assessment of mining process water quality. *Minerals*, 10, 940, 2020.
50. Le, T., Schreithofer, N., Dahl, O., Dissolution test protocol for estimating water quality changes in minerals processing plants operating with closed water circulation. *Minerals*, 10, 653, 2020.
51. Wiese, J., Harris, P., Bradshaw, D., The influence of the reagent suite on the flotation of ores from the merensky reef. *Miner. Eng.*, 18, 189–198, 2005.
52. Dzvinamurungu, T., Viljoen, K.S., Knoper, M.W., Mulaba-Bafubiandi, A., Geometallurgical characterisation of merensky reef and ug2 at the marikana mine, Bushveld Complex, South Africa. *Miner. Eng.*, 52, 74–81, 2013.
53. Sgs Minerals Services, *Pgm Ore Processing At Impala's Ug-2 Concentrator in Rustenburg, South Africa.*
54. Ekmekçi, Z., Bradshaw, D.J., Allison, S.A., Harris, P.J., Effects of frother type and froth height on the flotation behaviour of chromite in UG2 ore. *Miner. Eng.*, 16, 941–949, 2003.
55. du Bois, M., Giles, K.A., Hamilton, J.K., Rebers, P.A., Smith, F., Colorimetric method for determination of sugars and related substances. *Anal. Chem.*, 28, 350–356, 1956.
56. Wills, B.A. and Napier-Munn, T., *Mineral Processing Technology. An Introduction to the Practical Aspect of Ore Treatment and Mineral Recovery*, vol. 7th, Elsevier Science & Technology Books, 2006.
57. Kaushal, S.S., Likens, G.E., Pace, M.L., Utz, R.M., Haq, S., Gorman, J., Grese, M., Freshwater salinization syndrome on a continental scale. *Proc. Natl. Acad. Sci. U. S. A.*, 115, E574–E583, 2018.
58. Kaushal, S.S., Groffman, P.M., Likens, G.E., Belt, K.T., Stack, W.P., Kelly, V.R., Band, L.E., Fisher, G.T., Increased salinization of fresh water in the Northeastern United States, 2005.
59. *Fast and Accurate Thermo Scientific Gallery Discrete Industrial Analyzers.*
60. Stetson, S.J., Patton, C.J., Guaglione, N., Chestnut, Z., Measurement of nutrients in saline and hypersaline waters by discrete analyzer colorimetry without matrix matched calibration standards. *Talanta*, 203, 297–304, 2019.
61. Manono, M.S., Corin, K.C., Wiese, J.G., The effect of ionic strength of plant water on foam stability: A 2-phase flotation study. *Miner. Eng.*, 40, 2013.
62. Manono, M.S., Corin, K.C., Wiese, J.G., An investigation into the effect of various ions and their ionic strength on the flotation performance of a platinum bearing ore from the merensky reef. *Miner. Eng.*, 36–38, 2012.
63. Wiese, J., Harris, P., Bradshaw, D., The response of sulphide and gangue minerals in selected merensky ores to increased depressant dosages. *Miner. Eng.*, 20, 986–995, 2007.
64. Muzenda, E., Afolabi, A.S., Abdulkareem, A.S., Ntuli, F., Effect of PH on the recovery and grade of base metal sulphides (pgms) by flotation. *Proceedings*

of the World Congress on Engineering and Computer Science 2011, vol. II, pp. 19–22, 2011.

65. Kirjavainen, V., Schreithofer, N., Heiskanen, K., Effect of calcium and thiosulfate ions on flotation selectivity of nickel–copper ores. *Miner. Eng.*, 15, 1–5, 2002.
66. Manono, M., Corin, K., Wiese, J., The effect of the ionic strength of process water on the interaction of talc and CMC: Implications of recirculated water on floatable gangue depression. *Minerals*, 9, 2019.
67. Manono, M.S., Corin, K.C., Wiese, J.G., Inorganic electrolytes on the efficacy of a carboxymethyl cellulose as a coagulant for talc: Implications for talc depression in flotation, in: *Proceedings of the Mine Water – Technological and Ecological Challenges*, International Mine Water Association, Perm, Russia, Perm State University, pp. 217–224, 2019.
68. Laskowski, J.S., Liu, Q., O'Connor, C.T., Current understanding of the mechanism of polysaccharide adsorption at the mineral/aqueous solution interface. *Int. J. Miner. Process.*, 84, 59–68, 2007.
69. Dzingai, M., Manono, M.S., Corin, K.C., Probing the effect of water recycling on flotation through anion spiking using a low-grade CU–NI–PGM ore: The effect of NO3–, SO42– and S2O32–. *Minerals*, 11, 340, 2021.

6

Complexity of Chalcopyrite Mineral Affecting Copper Recovery During Leaching

Kolela J. Nyembwe[1], Frans Waanders[1], Martin Mkandawire[2], Bhekie Mamba[3] and Elvis Fosso-Kankeu[4]*

[1]Water Pollution Monitoring and Remediation Initiatives Research Group, School of Chemical and Minerals Engineering, Faculty of Engineering, North-West University, Potchefstroom, South Africa
[2]Department of Chemistry, School of Science and Technology, Cape Breton University, Nova Scotia, Canada
[3]Institute for Nanotechnology and Water Sustainability (iNano WS), College of Science Engineering and Technology (CSET), University of South Africa, Florida Science Campus, Johannesburg, South Africa
[4]Department of Metallurgy, School of Mining, Metallurgy and Chemical Engineering, Faculty of Engineering and The Built Environment, University of Johannesburg, Johannesburg, South Africa

Abstract

Regardless of the media/solvent used, chalcopyrite ($CuFeS_2$) is highly refractory and is considered to be one of the most difficult to leach. Its leaching is slow and incomplete owing to the diffusion barrier formed that prevents further attack of the leaching solution (solvent) on the unreacted $CuFeS_2$ mineral surface. The nature and composition of this barrier, together with the dissolution mechanism, remain controversial. There are three hypotheses for the structure of the formed diffusion barrier. The first hypothesis is that elemental sulfur (S^o), formed during leaching, prevents further diffusion of the reactant from reaching the un-leached chalcopyrite. The second hypothesis, commonly cited, ascribes the poor leaching to the formation of Cu-rich polysulfides, promoted by Fe preferential dissolution occurring through solid-state transformation (metal-deficient sulfide). The

**Corresponding author*: elvisfosso.ef@gmail.com

Elvis Fosso-Kankeu, Bhekie B. Mamba and Antoine F. Mulaba-Bafubiandi (eds.) Recovery of Values from Low-Grade and Complex Minerals: Development of Sustainable Processes, (145–178) © 2024 Scrivener Publishing LLC

third hypothesis blames iron-precipitates including jarosite, jarosite-like species, and goethite. This book chapter aims to explore the various challenges related to the mineralogy of chalcopyrite that affect the recovery of copper during the leaching process.

Keywords: Chalcopyrite, complex minerals, passivation, leaching, copper dissolution, hydrometallurgy

6.1 Introduction

The demand for copper (Cu) has risen during the last decade owing to the imbalance between its production and supply [1]. The imbalance is exacerbated by the depletion of available high-grade ore mineral that has led to increasing exploitation of mineral ores of higher complexity [2]. Consequently, Cu is increasingly extracted from the chalcopyrite ($CuFeS_2$), the most abundant Cu mineral in the world, but also known to be the most recalcitrant copper-sulfide mineral (Cu-S), when subjected to chemical or bioleaching processes [3–7].

Approximately 70% of metallic Cu is produced from the $CuFeS_2$ mineral, and over 90% of copper worldwide is obtained through the pyro-metallurgical route. This route uses thermal treatment (furnaces) for the chemical and physical transformation of material, which enables rapid recovery of the targeted metal. Though the pyro-metallurgical treatment has been relatively the most successful, this extraction method has a very high pollution footprint. Thus, the green production process of Cu is advocated, with the emphasis on minimizing pollution, including the need to avoid sulfur dioxide (SO_2) emissions and the market for sulfuric acid (H_2SO_4) as a secondary product from the green production process [2].

In addition, there is a relative decrease in profit margins for Cu mineral processing already caused by a scarcity of high-grade ore bodies [8]. The prolonged use of pyro-metallurgy has brought the need for the installation of new smelting operations able to treat low grade or complex ore bodies containing high impurity content and harmful electro-refining elements including be As, F, and Sb. Furthermore, there is a call for the construction of small leaching plants near mine sites rather than shipping the concentrate over long-range smelters, which will therefore reduce the freight and transport costs. Consequently, the hydrometallurgical route is appearing to be a more attractive way for Cu production due to its high level of selectivity and its viability from an environmental point of view.

Hydrometallurgy is the extraction of mineral from the ore by leaching with aqueous solvents, also known as the medium. It takes place by means of chemical reaction and can be characterized or categorized according to the media used to dissolve the mineral. $CuFeS_2$ can be dissolved in different media including sulfate, (H_2SO_4-$Fe_2(SO_4)_3$), chloride-based (HCl-$FeCl_3$), mixed chloride sulfate (H_2SO_4-$Fe_2(SO_4)_3$-NaCl) and nitrate-based ($Fe(NO_3)_3$) system. H_2SO_4-$Fe_2(SO_4)_3$ is the preferred dissolution medium for $CuFeS_2$ leaching because the system chemistry is simple and well understood, and the recovery of Cu from sulfate media by either solvent extraction or electrowinning is straightforward [9].

The hydrometallurgy process of copper can be conveniently subdivided into two main operations: the leaching and the recovery stages. The leaching or dissolution is the first step in the processing of metal extraction. In this stage, the targeted metal is selectively transferred from either naturally occurring minerals (copper ore) or matte to an active leach solution referred to as medium or solvent. After leaching, the solution undergoes the solid–liquid separation stage with the objective of rendering the solution suitable for purification and metal recovery. The concentration and purification of the desired metal are achieved at the solvent–extraction process stage, while the metal recovery is typically achieved at the electrowinning process step. It was observed that the quality of Cu directly produced by electrowinning from the sulfate system was similar to the one obtained through conventional smelting–refining, and also showed a much lower process cost. The principal interest of Cu hydrometallurgy is the development of a suitable economic and environmental-friendly sulfate-based leaching system for the $CuFeS_2$ mineral. Hydrometallurgical processing schemes virtually eliminate the production of gaseous pollutants, mainly sulfur dioxide (SO_2). They allow energy conservation and have been reported to be a more efficient and economical process route for the treatment of low-grade ore, complex ore matrix bodies, and concentrator tailings than the conventional smelter technologies [10, 11]. The main objective of this review is to highlight the various challenges encountered during the dissolution of chalcopyrite and discuss the principles of the parameters considered during the leaching process.

6.2 $CuFeS_2$ Crystal Structure

It was reported in an earlier study conducted by Burdick and Ellis [12] that the $CuFeS_2$ crystal structure was attributed to the zinc blende (ZnS) group. However, its tetragonal unit cell was found to be twice the volume of the zinc blende unit cell. In that structure, all Zn atoms were substituted

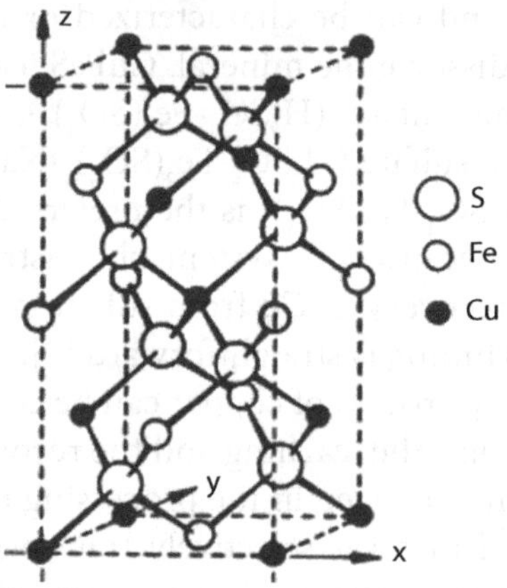

Figure 6.1 $CuFeS_2$ crystal structure.

by Cu and Fe, and the center of the tetrahedron was occupied by S, which was connected to two atoms of each metal (atom of Cu and two atoms of Fe). The $CuFeS_2$ unit cell is therefore composed of four copper (Cu), four iron (Fe), and eight sulfur (S) atoms as illustrated in Figure 6.1. The average bond length is 0.368, 0.225, and 0.230 nm individually for S–S, Fe–S and Cu–S, respectively. Owing to its metal excess, $CuFeS_2$ is a semiconductor of n-type at a band gap of 0.6 eV [2].

This structural configuration of the unit cell of chalcopyrite makes the mineral the most stable of the copper sulfide minerals, which, along with its abundance, promotes the extraction of copper from this mineral. The dimensional unit cell lattice constants of chalcopyrite are given as a = b = 5.2988 and c = 10.434 with the Cu–S, Fe–S, and S–S bond lengths being 0.230, 0.225, and 0.368 nm, respectively [13].

6.3 Application of Dissolution/Leaching of Chalcopyrite

Hydrometallurgy deals with the production of metal from the ore mineral by aqueous dissolution. The leaching or dissolution process is a vital operation in the hydrometallurgical treatment of minerals. As a result, its efficiency has a major impact on the various downstream processes and mainly on the economic and technical aspects of the mineral hydrometallurgy. Leaching involves the dissolution of the metal of interest from the mineral (ore) into the solvent, forming a solution containing the metal to be recovered. During leaching, various phenomena occur and may involve the precipitation, sorption, diffusion, and solid surface coating. The dissolution operation is mainly dependent on both the physical and chemical factors. While the important physical factors include the particle average size, its porosity, temperature, mineral phase homogeneity, time frame, and suspension of powdered solids, the chemical factors are mainly related to redox potential conditions and the concentration of hydrogen and hydrogen oxide that refer to the pH.

Different solvents (media) can be used to leach Cu from the $CuFeS_2$ mineral. These solvents include nitrates (NO_3^-), chlorides (HCl), ammonia (NH_3), and sulfates (SO_4^{2-}). Ferric sulfate (H_2SO_4-H_2O-$Fe_2(SO_4)_3$-$FeSO_4$) is the system of choice for the hydrometallurgy of $CuFeS_2$ mineral oxidation in sulfuric acid (H_2SO_4). It is also applied to chemical and bio-leaching dissolution in stirred-tank technologies for concentrate and heap or dump technologies for low-grade ores [8]. H_2SO_4-H_2O-$Fe_2(SO_4)_3$-$FeSO_4$ has received the most attention because it presents several advantages: Its chemistry is simple and easily handled; its low capital and operational costs; it is environmentally friendly; and was reported as being the most convenient and straightforward system for the recovery of copper by solvent extraction or electrowinning [9].

6.4 Challenges Associated with Copper Dissolution from the Chalcopyrite Mineral

Most researchers have agreed that regardless of the solvent used for leaching, $CuFeS_2$ is a highly refractory mineral, and its dissolution is slow and incomplete because of the formation of the passive film or diffusion barrier that progressively builds up on the mineral surface and hinders further solvent attachment on the undissolved $CuFeS_2$ core [14]. The composition of the diffusion barrier, together with the dissolution mechanism, remains subject to controversy [7, 15, 16].

$$CuFeS_2 + 2Fe_2(SO_4)_3 \rightarrow CuSO_4 + 5FeSO_4 + 2S^o \qquad (6.1)$$

In the ferric sulfate system, chalcopyrite dissolves according to Equation 6.1, with both sulfate (Cu and Fe) and elemental sulfur (S^o) as the reaction product. Three major hypotheses are used to explain the structure and chemistry of the dissolution barrier formed during $CuFeS_2$. The first hypothesis suggests that elemental sulfur (S^o), formed during leaching, prevents further diffusion of the reactants from reaching the un-leached chalcopyrite [17, 18], while the second points out the metal-deficient sulfide theory, which is the commonly cited reason explaining the $CuFeS_2$ slow leaching kinetics. It refers to the formation of Cu-rich polysulfides formed through the preferential dissolution of Fe, which promotes solid-state transformation [19, 20]. The last hypothesis proposes that iron-precipitates including jarosite, jarosite-like species, and goethite are the main constituents of the dissolution barrier transport process through and from the un-dissolved $CuFeS_2$ [20, 21].

6.4.1 Elemental Sulfur

During the dissolution of Cu from the $CuFeS_2$ mineral, sulfides are oxidized to elemental S^o. The formed sulfur layer surrounds the unreacted $CuFeS_2$ mineral surface and may, to some extent, influence the dissolution kinetics. Previous studies have reported that parabolic kinetics of $CuFeS_2$ in ferric sulfate are caused by the S^o formation phenomenon, and suggested that the dissolution mechanism was controlled by the diffusion process through a thick elemental sulfur layer [10, 22, 23]. The chemistry of elemental sulfur is very complex. It is, therefore, logical to consider all the possibilities concerning the nature and evolution of sulfur and its different existing allotropic modifications [23].

Previous studies have reported that the dissolution of $CuFeS_2$ proceeds through the formation of non-stoichiometric polysulfide layers [20]. These new sulfidic species were seen to have slower leaching kinetics than the original $CuFeS_2$ surface, and therefore, they hindered copper dissolution [20]. It was reported that the formation of S^o was responsible for the parabolic kinetics observed, suggesting that the initial dissolution stage had been strictly controlled by the chemical reaction and, over time, it had changed to the control of the diffusion rate owing to the compaction of sulfur layers. It further suggested that leaching conditions should allow rapid removal of S^o from the un-leached $CuFeS_2$ mineral [11, 14, 24–27].

However, further research has ruled out that S^o is responsible for the slow, incomplete, and parabolic behavior of $CuFeS_2$ dissolution. It was observed that S^o had a porous morphology, which allowed transport processes through and from the dissolving mineral. Therefore, it could not be held responsible for passivation [28]. In addition, it was identified that removing the sulfur coating from the un-leached mineral did not improve Cu dissolution [29]. The structure of S^o also seemed to depend on pH as porous S^o was observed in slightly acidic media, which were also characterized by a rapid dissolution rate of $CuFeS_2$ and high recovery of copper, whereas, in the case of a highly acidic solution, a dense S^o is formed and leads to a weak dissolution of copper [30–32].

6.4.2 Iron Precipitates

Various iron precipitates are also susceptible to hindering further mineral dissolution. These precipitates are formed owing to Fe^{3+} hydrolysis and include compounds such as jarosite, jarosite-like species, and goethite. Their formation was reported to retard $CuFeS_2$ dissolution by restricting and preventing the mass transfer of ions and access to the mineral surface [23]. These precipitates were also found to be temperature- (T), acid level- (pH), and oxido-reduction potential- (ORP) dependent. Equation 6.2 and Equation 6.3 summarize the formation of jarosite and goethite, respectively.

$$Fe_2(SO_4)_3 + (Na, K, H_3O)_2SO_4 + 12\ H_2O \rightarrow (Na, K, H_3O)\ Fe_3(SO_4)_2\ (OH)_6 \downarrow + 6\ H_2SO_4 \tag{6.2}$$

$$Fe_2(SO_4)_3 + 4H_2O \rightarrow 2FeOOH + 3H_2SO_4 \tag{6.3}$$

Most researchers have identified the presence of jarosite and/or goethite in the leached solid tailings, and have suggested that these iron products could be passive layer components that prevent further dissolution of Cu from $CuFeS_2$ during its leaching process [25, 32, 33]. Linear kinetics were observed at the early stage of dissolution, which was reported to have changed to parabolic rate as the dissolution progressed, as a result of the formation and precipitation of jarosite at the surface of the mineral [14].

Another study suggested that it is uncertain whether jarosite alone is responsible for the slow dissolution kinetics observed for the mineral $CuFeS_2$ [2]. Iron by-products have been observed to have little effect on the rate of Cu dissolution and recovery. The partial removal (70%) of jarosite by bio-reduction of the ferric ion revealed that the dissolution kinetics and recycling of copper were not significantly improved [34].

6.4.3 Fe-Deficient Polysulfide

Based on its conductive property, the $CuFeS_2$ mineral has a resistivity between $2x10^{-4}$ and $9x10^{-3}$, and its dissolution in an electrolyte can be interpreted electrochemically. Two half reactions summarize oxidation and reduction (Equation 6.4 and Equation 6.5, respectively), which take place during the $CuFeS_2$ leaching process.

$$\text{Anodic: } CuFeS_2 \rightarrow Cu^{+2} + Fe^{+2} + 2S + 4e- \quad (6.4)$$

$$\text{Cathodic: } 4Fe^{+3} + 4e- \rightarrow 4Fe^{+2} \quad (6.5)$$

Both reactions, Equation 6.4 and Equation 6.5, occur concurrently on the dissolving mineral surface at a rate that ensures that the net sum of electrons produced is null. This condition (Σ é=0) is met at a specific potential referred to as the mixed potential. Owing to the fact that both the cathodic and anodic reactions are independent, they can be separated and individually studied, and their rate expressions established according to that. The net potential at which the corrosion takes place is obtained through the combination of both anodic and cathodic potential currents on the Evans diagrams. Electrochemical studies of $CuFeS_2$ dissolution have gained a great deal of interest because they can provide beneficial information concerning the nature of passive film, dissolution mechanism, and kinetics [2].

Several studies have identified the presence of a transitional phase believed to be a metal-deficient formation, which appears to depend mainly on the potential [29]. The formation of this phase was promoted by solid-state transformation owing to fast preferential dissolution or withdrawal of Fe over Cu [19, 35]. It is believed that the metal-deficient phase promotes the slow reduction rate of Fe^{3+} on the $CuFeS_2$ mineral surface, owing to the slow diffusion of the ions, electron transfer, and the supply of holes to the solid-solution interface.

This metal-deficient polysulfide appeared to be a rich copper polysulfide of the form of ($Cu_{1-x}Fe_{1-y}S_{2-z}$). The formed intermediate could further decompose and form a second (non-stoichiometric) copper sulfide phase $CuS_{(n-s)}$ at a potential of between 700 and 750 mV [19], according to Equation 6.6:

$$Cu_{1-x}Fe_{1-y}S_{2-z} \rightarrow (y\text{-}x)\ CuS(_{n-s}) + (y\text{-}x)\ CuS(_{n-s}) + (1\text{-}y)\ Cu^{2+}$$
$$+ (1\text{-}y)Fe^{2+} + (2+x+y)\ S^o + 4(1\text{-}y)e^- \quad (6.6)$$

$CuFeS_2$ dissolution has been explained according to different stages based on the occurrence of the metal-deficient phase. A two-stage dissolution process was proposed for the anodic dissolution in which the release of Cu^{2+} related to the first stage while the second stage was the dissolution of Fe^{2+} according to the sequence reaction, which could lead to the formation of an unstable polysulfide phase [2]. The three-step process includes the dissolution, passive and trans-passive region. From the anodic dissolution of the $CuFeS_2$ mineral, two main intermediate phases appeared to change in sequence chalcopyrite→bornite→covellite. Both these intermediates could also be categorized as metal-deficient polysulfide owing to their decrease in Fe content.

$$n\,Cu_{1-x}Fe_{1-y}S_{2-z} \rightarrow Cu_{n-1}Fe_{n-1}S_{2n} + Cu^{2+} + 2e^{-} \tag{6.7}$$

$$Cu_{n-1}Fe_{n-1}S_{2n} \rightarrow Fe^{3+} + Cu_{n-1}Fe_{n-1}S_{2n} + 3e^{-} \tag{6.8}$$

The overall reaction can be expressed by:

$$CuFeS_2 \rightarrow Fe^{3+} + Cu^{2+} + Cu_{n-1}Fe_{n-1}S_{2n} + 5e^{-} \tag{6.9}$$

6.4.4 Gangue-Related Mineral

Recently, it has been reported that gangue-related minerals belonging to carbonates, such as calcite $CaCO_3$ and/or dolomite $CaMg(CO_3)_2$, can also retard the dissolution of Cu from the $CuFeS_2$ mineral. An XRD analysis of solids-leached residues revealed the presence of gypsum ($CaSO_4$-H_2O) precipitates, which could have probably coated the undissolved $CuFeS_2$ mineral [36, 37]. This could suggest that the formation of $CaSO_4$-H_2O had prevented further $CuFeS_2$ mineral oxidation, restricting the exposure of the mineral surface to the lixiviant. It was reported that a sufficient quantity of carbonate in the mineral matrix promotes acid consumption [38]. In addition to that, the acid-consumption phenomena generate the formation and the release of species that are able to cause the formation of by-products and precipitates including jarosite and aluminite [39].

6.5 H_2SO_4-$Fe_2(SO_4)_3$-$FeSO_4$-H_2O Speciation

Acidic ferric sulfate (H_2SO_4-$Fe_2(SO_4)_3$-$FeSO_4$-H_2O) is usually prepared by mixing known amounts of iron sulfate ($Fe_2(SO_4)_3$) and/or $FeSO_4$, sulfuric acid (H_2SO_4), and water (H_2O)). In this system, Fe is distributed as free

Fe(III) and Fe(II) or as complex iron compounds (e.g., $FeHSO_4^+$, $FeSO_4^o$, $Fe(HSO_4)^{2+}$, $Fe(SO_4)_2^-$, and $FeSO_4^+$). The content of these various species in solution mainly depends on the pH and temperature [40–42]. Figure 6.2 summarizes the effect the solution/media pH has on Fe(III) and Fe(II) speciation at 25°C. From the figure, it can be seen that the majority of Fe(III) species are present in the form of $Fe(SO_4)_2^-$ complexes and $FeSO_4^+$ within the pH range of 1 to 2. This clearly shows that the leaching kinetics may be correlated to the distribution of different iron species existing in solution. Therefore, the oxidizing power/strength of H_2SO_4-$Fe_2(SO_4)_3$-$FeSO_4$-H_2O solution is revealed as a function of all the various existing ferric species solutions.

Hirato and co-authors [43] examined Fe speciation of an acidified ferric sulfate solution ($Fe_2(SO_4)_3$-H_2SO_4-H_2O) and plotted the distribution Fe species, according to Sapieszko and co-authors [44], and concluded that the Fe^{3+} and $FeHSO_4^{2+}$ are the most important Fe species for the dissolution of $CuFeS_2$ in the media.

Hiroyoshi and co-authors [45] reported that Fe(II) ions were more effective for the dissolution of $CuFeS_2$ than Fe(III). From their study, higher Cu recoveries were obtained in the presence of adequate quantities of ferrous ions (Fe^{2+}) than in the case where Fe^{2+} was not present. Their results revealed further that Fe^{2+} alone was not responsible for the Cu withdrawal from the $CuFeS_2$ mineral, but that O_2 played a vital role for Cu dissolution. Hirato and co-authors [43] observed a drop in Fe^{3+} and $Fe(HSO_4)^{2+}$ related

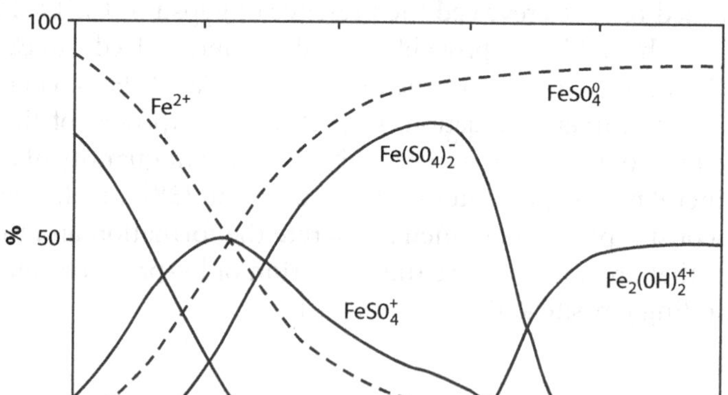

Figure 6.2 Effect of pH on the Fe speciation at 25°C.

to the Cu dissolution rate in their study. They concluded that both Fe^{3+} and $Fe(HSO_4)^{2+}$ ions had a significant role to play during the dissolution of chalcopyrite.

6.6 Parameters Affecting Dissolution

Although the formation and mechanism of the passivation film are still disputed and no general consensus has been reached in many studies on the phenomenon, various investigations have effectively identified certain dissolution parameters that are known to influence and/or improve the leaching kinetics of copper from the chalcopyrite host mineral. These dissolution parameters include the acidity of the solution (pH), leaching temperature (T (°C)), particle size distribution, mineralogical composition, redox potential (Eh), oxidant concentration, and lastly, addition of additives [31, 46].

6.6.1 Effect of pH

Previous studies have pointed out that the media pH has a poor response to the dissolution rate of $CuFeS_2$ in acidified ferric sulfate [10, 47]. In this regard, a low pH value must be maintained in order to keep iron in the solution and to avoid its hydrolysis and/or the formation of hydroxy precipitates [48, 49]. These precipitates include $Fe(OH)_3$ and other ferric hydrosulfates, goethite, and jarosite, and are able to restrict mass transfer and prevent Fe^{3+} (oxidant) attack on the un-leached mineral surface.

The effect of acidity also appeared to depend on the dissolution temperature. It was observed that the level of acidity had little effect at a low leaching temperature (25°C) compared with a high temperature (68°C), which interfered with the oxidation process by preventing further copper dissolution from the chalcopyrite mineral during electrochemical experiments [47]. However, an increase in the dissolution rate and recovery of Cu has been reported to be because a decrease in the pH of the solution from 3 to 1 [16, 31]. Contrary to that, at 68°C, rapid rate accompanied by high copper recoveries have been reported during the chemical leaching of chalcopyrite in ferric sulfate at pH values varying between 1.5 and 2.0 as opposed to low pH values: 0.5 and 1.0. However, a high pH also revealed the highest content of precipitates of iron compounds (goethite and jarosite) in their corresponding solid residues [32].

The oxidation of $CuFeS_2$ for Cu dissolution appeared feasible at relatively high pH values in the presence of O_2 [46], and an earlier study highlighted that the extraction of Cu increased with the increase in the acid content of up to 27 g/L H_2SO_4. Over and above this, valuable copper extraction was very low [30]. This tends to support the practical acidity of ferric solution usually maintained between 1.0 and 1.8 [2].

6.6.2 Effect of Size Particle

Particle size is also a vital and essential parameter for dissolution as it is directly related to mineral exposure. Previous studies have reported a rapid dissolution rate and recovery with increasing particle area [10]. Three particle fractions comprising -74, +53, -53, +38, and -38 μm were studied for their kinetic dissolution of copper from the mineral chalcopyrite. The results revealed that in order to achieve the best copper extraction within a reasonable timeframe, chalcopyrite would necessarily be grounded below 75 μm [50]. In addition, almost complete dissolution of copper with a rapid rate was observed when dissolving a particle size of <10 μm, while a coarse particle size of +20 to -106 μm revealed a low dissolution rate and only 32% recovery after 96 hours [51].

The significant improvement in the dissolution of $CuFeS_2$ could be linked to the mineral exposure surface favored by an operation linked to fine grinding. The latter refers to the mineral mechanical chemical activation under which the physicochemical and structural changes are induced by the deformation or transformation responsible for the crystal defect of the mineral leading to an improved dissolution kinetics [52, 53].

6.6.3 Effect of Concentration

In general, most chemical reactions proceed more quickly when the concentration of one or more reagents is increased [54]. It is therefore logical to study the effect of the oxidant (Fe^{3+}) on the speed and recovery of Cu during the dissolution of $CuFeS_2$ in ferric sulfate. The previous investigation showed a first order dissolution rate of $CuFeS_2$ with respect to the concentration of $Fe_2(SO_4)_3$ up to 0.1 mol dm^{-3} and a further increase in the concentration of $Fe_2(SO_4)_3$ had a negligible response on leaching [16].

A study of a previous work, accompanied by an experimental laboratory evaluation to assess the effect of Fe^{3+} on the dissolution rate of $CuFeS_2$, showed that the dissolution of $CuFeS_2$ in ferric sulfate was relatively independent of the concentration of Fe^{3+} [48]. In addition, it has been reported that the concentration of iron (III) has little impact and influence on the dissolution rate of Cu from the mineral $CuFeS_2$ [46].

6.6.4 Effect of Temperature

Raising the temperature reveals a positive response to the rate of metal dissolution and recovery for most chemical reactions. More than 10 times the extraction of Cu have been identified when dissolving $CuFeS_2$ by raising the temperature from 30°C to 90°C [10, 20, 43, 48, 55].

The other kinetic parameter associated with the effect of temperature and reactant concentration is the activation energy (Ea), which represents the minimum amount of energy necessary to activate the different species in reaction to a condition where they can undergo a chemical transformation or physical transport. Its value gives an overview of the driving force of the reaction mechanism, be it mass transfer ($\leq$ 20 kJ.mol^{-1}), mixed control (20 $\leq$ Ea $\leq$ 40 kJ.mol^{-1}), or the reaction of surface 40 kJ.mol^{-1} $\leq$ Ea) [2].

The dissolution of $CuFeS_2$ is characterized by a high activation energy value ranging from 70 to 88 kJ.mol^{-1} [10, 43]. Similar dissolution rates and activation energy (Ea = 70 kJ.mol^{-1}) were reported for 11 different chalcopyrite samples [3] dissolved in ferric sulfate. However, a low activation energy (Ea = 47 kJ.mol^{-1}) was estimated for the dissolution of $CuFeS_2$ in Fe_2 $(SO_4)_3$-H_2SO_4 through tight control of Fe^{3+}/Fe^{2+} by careful addition of $KMnO_4$ [56], while an activation energy of 83 kJ.mol^{-1} has been reported upon the dissolution of $CuFeS_2$ in H_2SO_4 under the presence of nitrate ions (NO_3^{2-}). Elsewhere, an appreciably high Ea value (Ea = 103 kJ.mol^{-1}) for the dissolution of $CuFeS_2$ in the $Fe_2(SO_4)_3$-H_2SO_4 system was also reported.

It should be noted that the different values of Ea reported in the literature vary depending on the method used to calculate the rate. Ea is usually determined by the fit of the data, and if the data are scattered, the resulting activation energy value would be different. In addition, it was also observed that the activation energy varied depending on the dissolution medium. Chloride media have been reported to have a much lower Ea than sulfate systems. The particle size also has an influence on the estimated activation energy. Fine particles appeared to possess low Ea values. The kinetics of dissolution of $CuFeS_2$ in an alkaline glycine or glycinate medium for two particles of different sizes, 20–38 μm and ultra-fine size fraction, were studied; the results revealed Ea values of 71.9 kJ.mol^{-1} and 29.6 kJ.mol^{-1}, respectively [51].

6.6.5 Effect of Potential

The potential of the solution also plays an essential role during the dissolution of $CuFeS_2$. The literature reports the existence of a narrow potential region (400–450 mV/Ag/AgCl) in which the yield and dissolving rate of Cu are high. While at redox potential values above this range, the dissolution is hampered owing to the blockage at the mineral surface by the precipitation of jarosite [45]. It was pointed out that at this potential (400–450 mV), the ferric-to-ferrous (Fe^{3+}/Fe^{2+}) ratio seemed to reach equilibrium, contrary to high potential values, which would favor the rapid precipitation of the Fe^{3+} products, leading to $CuFeS_2$ passivation [4].

This could suggest that the active state of chalcopyrite is maintained at low potential (440 mV Ag/AgCl), while at moderate potential values (between 440 and 510 mV), $CuFeS_2$ is bistable, suggesting that the mineral could passivate or remain active depending on how it has been brought to this potential. Above 510 mV Ag/AgCl, a slow dissolution rate linked to passivation is observed [57].

High copper recoveries have been reported from various studies under low potential conditions associated with high values; more than 80% of copper was obtained at 410 mV Ag/AgCl [57], while $CuFeS_2$ passivation occurred at 600 mV Ag/AgCl [58]. A rapid dissolution rate of Cu was also observed from the leaching of $CuFeS_2$ at 420 mV Ag/AgCl [59]. It appears that monitoring and control of the solution potential represent a key factor for increasing dissolution rates in $CuFeS_2$ leaching systems. The majority of previous research apparently converged to agree with the existence of a narrow potential range 400–450 mV (Ag|AgCl) where both the dissolution rate and yield are at the maximum.

6.6.6 Effect of Geology of the Host Ore Body

The mineral $CuFeS_2$ crystallizes in different host ores, so it could be accompanied by various different mineral species, including carbonatitic and siliciclastic, which could impact on the leaching behavior of the mineral $CuFeS_2$ in solution.

A typical example of carbonatitic chalcopyrite is the carbonatite chalcopyrite ore from Phalaborwa in South Africa (SA). This carbonatitic complex contains enough copper in the form of sulfides to constitute an economical deposit for the production of copper. The ore assemblage is igneous in nature, in which the mineral $CuFeS_2$ is accompanied by calcite ($CaCO_3$) or dolomite ($CaMg(CO_3)_2$) with magnetite (Fe_3O_4), traces of iron sulfides pyrite (FeS_2) and pyrrhotite ($Fe(1\text{-}x)S$) [60]. On the other hand, siliciclastic chalcopyrite could be identified in the region of the sedimentary copper belt of Central Africa. This region extends over both the Democratic Republic of Congo (DRC) and Zambia and is hosted by the Neoproterozoic rock of the Katanga Province (DRC). The host rock combines schist, quartzite in the form of arkose and dolomite [61]. It also contains sulfides (Cu-Co-Fe), in particular, chalcopyrite, pyrite, cobaltiferous pyrite, bornite, and carrolite. The dissolution response and rates of the $CuFeS_2$ could also be affected by its mineralogical origin and genesis. Variations in the leaching rate have been caused by the presence of impurities. For instance, high-iron content in sphalerite (Fe-ZnS), when present during the chalcopyrite ($CuFeS_2$) leaching, promotes the retarding effect on Cu dissolution, but low-iron sphalerites behave erratically.

This supports the galvanic corrosion mechanism except for the observed effects of molybdenite. The presence of pyrite leads to enhanced Cu dissolution owing to $CuFeS_2$-FeS_2 galvanic couple, with a high content of acid-consuming mineral that could lead to the release of species that are able to cause the formation of by-product precipitates, including jarosite and aluminite [62], which could hinder high Cu recoveries.

6.6.7 Effect of Additives

6.6.7.1 Addition of Silver (Ag^+)

Silver (Ag) ions have positively enhanced copper dissolution from $CuFeS_2$ leaching. It appears that Ag+ increases mineral reactivity and inhibits the hydrolysis of Fe^{3+} products on the mineral surface. For example, the dissolution of $CuFeS_2$ under O_2 leaching yielded 51% of Cu at 100°C for 30 min, while, under the same conditions, with the addition of a low content of Ag^+ (0.75 %), the dissolution of $CuFeS_2$ yielded 95 % of copper.

Ag^+ addition leads to the formation of Ag_2S (Equation 6.10), which promotes a high recovery of Cu during $CuFeS_2$ dissolution as the morphology of the S^o formed appears porous and allows electrons transport at the $CuFeS_2$ mineral surface [62].

$$CuFeS_2 + 4Ag^+ \rightarrow 2Ag_2S + Cu^{2+} + Fe^{2+} \quad (6.10)$$

Up until now, the Ag catalyzed effect on the $CuFeS_2$ dissolution process has not been commercialized owing to its high operating redox potential, which could lead to Fe^{3+} precipitation in the form of argentojarosite [18]. Once a product has formed, it is difficult to recover it as it contains Ag. In their study, Salehi *et al.* [18] opted for the dissolution of $CuFeS_2$ in the presence of FeS_2 particles coated by Ag; the leaching process was very fast and a relatively complete copper extraction was achieved within about 10 hours.

6.6.7.2 *Addition of Chloride (Cl^-)*

The presence of chloride ions in a sufate-based system has also been reported to enhance Cu dissolution from the $CuFeS_2$ mineral. The action of Cl^- during $CuFeS_2$ dissolution also appeared to change the morphology of the tight and tenuous sulfur layer, it having formed a much more porous film, which allowed further dissolution of $CuFeS_2$ at a reasonable rate, and this could be clearly noticed during the electrochemical dissolution using both the transient and steady-state techniques [63].

An increase in copper yield has been reported after the addition of chloride ions [64, 65]. A recovery from 40% to 90% was reported after the addition of 2 M of NaCl during $CuFeS_2$ dissolution (10 hours) of $CuFeS_2$ in acidified ferric sulfate (50 g/L) at 95°C [66]. Furthermore, the solid leached residues displayed high surface porosity, which could suggest that the presence of Cl^- ions promotes the reagents and products diffusion through the porous S^o formed, leading to high Cu recovery. Nicol *et al.* [64] and Liu *et al.* [65] also observed the positive effect of Cl^- ions on Cu withdrawal from the $CuFeS_2$ dissolution.

It is worthwhile to highlight that in the presence of Cl^- ions and under oxidative dissolution of $CuFeS_2$, copper is more than likely to generate complexes including $CuCl_3^{2-}$, $CuCl^o$, and $CuCl_2^-$, which have a negative effect on the further dissolution of Cu. These complexes further inhibit Cu withdrawal from the mineral.

6.7 Thermodynamic Considerations

The $CuFeS_2$ dissolution is usually discussed with the aid of the Pourbaix diagram (E-pH) (Figure 6.3). The diagram serves as initial identification for eventual dissolution reactions under both oxidizing and reducing environment at different pH values. From the diagram, it can be observed that in acidic media, the increase in potential value on the surface of the dissolving mineral leads to the occurrence of different oxidation reactions, which favor the formation of intermediate phases. These transitory phases are Cu-rich minerals and include the formation of bornite (Cu_5FeS_4), chalcocite (Cu_2S), and covellite (CuS) from which successful Cu dissolution can be achieved at potential above 0.5 V at pH values ranging from 0 to 2.

$$CuFeS_2 + Fe_2(SO_4)_3 \rightleftharpoons CuS + 3FeSO_4 + S \quad \Delta G = -18.6 \text{ (kcal)} \tag{6.11}$$

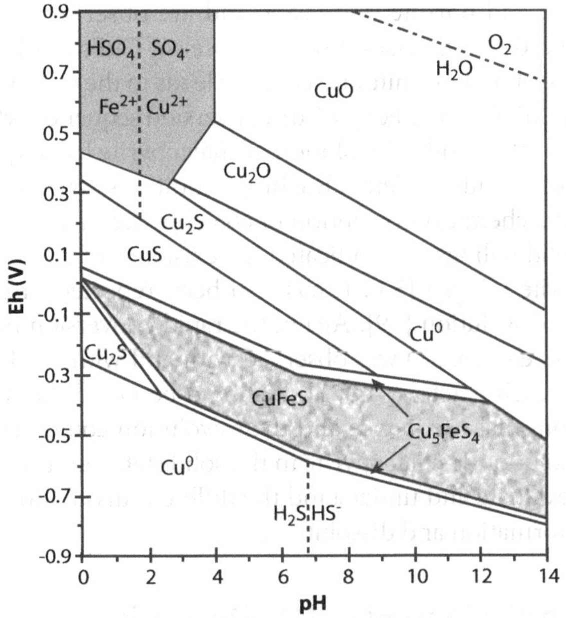

Figure 6.3 Potential-pH diagram of the Cu-Fe-S-H_2O system at 25°C: all solutes at 0.1-M activity except Cu.

$$2CuFeS_2 + 2Fe_2(SO_4)_3 \rightleftharpoons Cu_2S + 6FeSO_4 + 3S \quad \Delta G = -30.8 \text{ (kcal)} \tag{6.12}$$

$$5CuFeS_2 + 4Fe_2(SO_4)_3 \rightleftharpoons Cu_5FeS_4 + 12FeSO_4 + 6S \quad \Delta G = -68.7 \text{ (kcal)} \tag{6.13}$$

6.8 $CuFeS_2$ Phases Conversion/Copper Sulfide (Cu-S) Intermediate Phases

Mineral alteration, transformation, and replacement are known to occur as a result of chemical reactions. This phenomenon has also been identified during mineral dissolution by which the dissolving mineral passes through intermediate phases before it is completely dissolved [67]. Various techniques have been used to identify intermediate or transitional phases, and previous studies have reported their presence during the chemical dissolution of chalcopyrite. These transient phases include chalcocite (Cu_2S), covellite (CuS), and bornite (Cu_5FeS_4), and are observed under both the oxidation and reduction dissolution processes of $CuFeS_2$ [67].

The presence of both bornite and covellite leads to the two-step dissolution process with a $CuFeS_2 \rightarrow Cu_5FeS_4 \rightarrow CuS$ conversion sequence, which has been proposed during the anodic dissolution of chalcopyrite [68, 69]. Covellite has also been reported under different leaching conditions. Its presence was identified during the chemical dissolution of chalcopyrite in acidic ferric sulfate. In addition, solid tailings also indicated the existence of another sulfide rich in copper, geerite (Cu_8S_5) [55]. It has also been reported in the solid residues of anodic dissolution [69]. Again, the rapid conversion of chalcopyrite to covellite ($CuFeS_2 \rightarrow CuS$) was observed with and without the presence of oxygen in the leach reactor [70, 71]. It should be noted that the formation of the transition-phase sequence and their evolution could help explain the process of dissolution of chalcopyrite in the solid state. For this, it is therefore important to evaluate and understand the different dissolution factors likely to affect their formation and dissolution.

6.8.1 Alternative Ways of $CuFeS_2$ Dissolution

Galvanic interactions represent the approach to enhance sulfide dissolution [72]. These interactions occur when conducting materials (minerals or metal) that are in close contact in a medium, which allows charge/electrons (e^-) transfer or flow. The inherent and appreciable electrical conductivity of most sulfide minerals mainly relies on their covalent bond characters, which provide non-localization of charges. Table 6.1 summarizes the electronic, structural, and resistivity of $CuFeS_2$, Fe_3O_4, and FeS_2. It shows that the galvanic interaction revealed a positive impact on the mineral dissolution. It appeared that the interaction modifies the cathodic and anodic half-reaction rates at the surface of each mineral [73]. The presence of active grain, such as pyrite (FeS_2), in the chalcopyrite deposit forms a galvanic pyrite–chalcopyrite (FeS_2-$CuFeS_2$) couple, which improves the dissolution of $CuFeS_2$ (anode), while FeS_2 (cathode) becomes galvanically protected.

A particular interest was placed on this galvanic couple, which led to numerous experimental works, all with the aim of understanding and providing insight into this alternative way of dissolving $CuFeS_2$. It has been reported that the dissolution of $CuFeS_2$ (anode) is significantly improved. Fast speed and high efficiency have been observed owing to the presence of FeS_2 [74–76]. In the absence of the galvanic couple with another mineral, both reactions take place on the surface of the mineral $CuFeS_2$. Figure 6.4 graphically summarizes the electrochemical oxidation of $CuFeS_2$ in $Fe_2(SO_4)_3$-H_2SO_4

Table 6.1 Mineral characteristics.

Name	Formula	Resistivity	Conductor	Structure	Ionic structure
Chalcopyrite	$CuFeS_2$	2.10^{-4} - 9.10^{-3}	n	Tetragonal	$Cu^+Fe^{3+}(S^{2-})_2$
Magnetite	Fe_3O_4	2.10^{-4} – 4.10^{-5}	n and p	Cubic	$Fe^{3+}[Fe^{3+}Fe^{2+}](O^{2/})_4$
Pyrite	FeS_2	3.10^{-2} - 1.10^{-3}	n and p	Cubic	$Fe^{2+}S^{2-}{}_2$

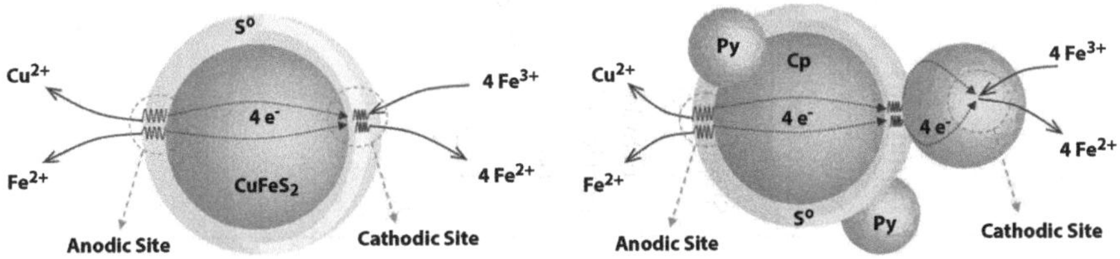

Figure 6.4 Schematic diagram of typical electrochemical leaching [74].

without (a) and with (b) the presence of FeS_2. The mineral characteristics are depicted in Table 6.1.

The electrochemical dissolution studies have revealed that the slow $CuFeS_2$ dissolution rate could be attributed to the anodic half-cell reaction [74, 75], which appeared to be promoted by the slow cathodic half-cell reaction [2]. Appreciable leaching kinetics have been obtained through the Galvaonx™ process from which 98% of copper is recovered during chalcopyrite dissolution in the sulfate system. The process makes use of the pyrite/chalcopyrite galvanic couple, through a mixed feed at ration of 2:1 to 4:1 for pyrite against chalcopyrite, respectively. Effective copper dissolution was obtained at a solution potential varying from 440 to 470 mV Ag/AgCl and temperature ranging between 70°C and 80°C [50, 74].

Magnetite (Fe_3O_4) could also be found as an existing impurity in the $CuFeS_2$ ore matrix, especially in the case of carbonatite [77]. Being a spinel, magnetite ($Fe^{2+}Fe^{3+}2O_4{}^{2-}$) in aqueous medium could represent a substantial source of oxidant in the system. However, it has been reported that Fe_3O_4 also forms a galvanic couple when combined with $CuFeS_2$, and could represent another route for the dissolution of copper from the mineral $CuFeS_2$ [72]. It is therefore of interest to conduct an additional dissolution study in order to establish operational conditions, which allow maximum dissolution of copper with respect to solution pH, particle size, redox potential, and temperature.

Table 6.2 Summary of some of the commercial processes.

Lixiviant type (Sulfate or chloride)	Name of the commercial process and country	Specific characteristics (temp, Pres…)	Cu recovery	Equations	Observations
Sulfate	Intec Mt Gordon Copper Process (Qld, Australia)	77°C–90°C with low-pressure autoclave oxidation	With chalcocite ores, the plant capacity achieved 50,000 tons Cu (LME grade) early in 2002	Stage 1 Cu2S + Fe2(SO4)3 → CuSO4 + CuS + 2 FeSO4 Stage 2 CuS + Fe2(SO4)3 → CuSO4 + S + 2 FeSO4 Ferric sulfate regeneration: 4 FeSO4 + O2 + 2 H2SO4 → 2 Fe2(SO4)3 + 2 H2O	It is reported as the lowest plant-operating cost in the world (cost for ore through metal was ~$0.31/lb Cu).
Sulfate	Phelps Dodge Total Pressure Oxidation (Baghdad Plant, Az, USA)	~225°C, and high pressure for successful minerals oxidation in sulfuric acid (H_2SO_4), then SX/EW	16,000 ton per annum Cu is recovered from concentrations	/	Iron precipitation takes place as hematite

(*Continued*)

Table 6.2 Summary of some of the commercial processes. (*Continued*)

Lixiviant type (Sulfate or chloride)	Name of the commercial process and country	Specific characteristics (temp, Pres…)	Cu recovery	Equations	Observations
Sulfate	Sepon Copper Process (Laos)	Atmospheric conditions	Initial capacity of 60,000 megatons per annum of copper	/	The leached residues are washed and floated. Autoclave treatment of flotation concentrate producing a basic ferric sulfate product.
Chloride	Outokumpu Hydrocopper Process (Pori, Finland)	The combination of air O_2 and chlorine Cl- while the temperature is kept at 85°C–90°C.	/	$CuFeS_2 + CuCl_2 + 3/4O_2 \rightarrow 2CuCl + 1/2Fe_2O_3 + 2S$ $CuCl + 1/2Cl2 \rightarrow CuCl_2$	After leaching, Cl- oxidizes a part of cuprous (Cu+) to form cupric chloride ($CuCl_2$)

(*Continued*)

Table 6.2 Summary of some of the commercial processes. (*Continued*)

Lixiviant type (Sulfate or chloride)	Name of the commercial process and country	Specific characteristics (temp, Pres…)	Cu recovery	Equations	Observations
Sulfate	Anglo American/ UBC Process	Fine ground chalcopyrite concentrates are treated under mild O2 pressure and at a temperature of 150 °C in an acid-sulfate system	Cu and Zn extractions >95%, elemental sulfur 60% in 2-hour residence time. Au extraction > 80%.	/	Surfactant added to disperse molten sulfur. Copper recovered by SX/EW Hematite/ jarosite formation.
Sodium Nitrite	Processing, Butte, Montana)	Moderate-pressure oxygen 30 min leach (125°C–175°C) of ultrafine ground concentrate.	Cu yields >99%	/	Commercially proven for argentiferous-tetrahedrite ores containing chalcocite.

(*Continued*)

Table 6.2 Summary of some of the commercial processes. (*Continued*)

Lixiviant type (Sulfate or chloride)	Name of the commercial process and country	Specific characteristics (temp, Pres...)	Cu recovery	Equations	Observations
Sulfate	Activox (Western Minerals Technologies, Perth WA)	Low-pressure oxygen sulfate leaching (~100°C) of ultrafine ground concentrate (P80=7–10 μm)	/	/	Demonstrated for mixed nickel, cobalt, and copper-bearing feeds in Africa. Successfully tested with chalcopyrite, demonstration plant in Botswana built.

(*Continued*)

Table 6.2 Summary of some of the commercial processes. (*Continued*)

Lixiviant type (Sulfate or chloride)	Name of the commercial process and country	Specific characteristics (temp, Pres…)	Cu recovery	Equations	Observations
Sulfate-Chloride	CESL (Cominco, Vancouver, Canada)	Moderate temperature (150°C) and high-pressure of light-ground chalcopyrite-chalcocite concentrate.	/	$3CuFeS_2 + 7.5O2 + H_2O + H_2SO_4 \rightarrow CuSO_4{\cdot}2Cu(OH)2 + 1.5Fe_2\ O3 + 6So$ $CuSO4.2Cu(OH)2(s) + 2H_2SO_4 \rightarrow 3CuSO_4(aq) + 4H_2O$	High-cost reagents and use of Ti autoclave apparatus necessary. Pilot-plant agreement with Companhia Vale do Rio Doce (CVRD) to operate on feed stock from Alamoa and Salobo, Brazil.

(*Continued*)

Table 6.2 Summary of some of the commercial processes. (*Continued*)

Lixiviant type (Sulfate or chloride)	Name of the commercial process and country	Specific characteristics (temp, Pres…)	Cu recovery	Equations	Observations
Sulfate	Dynatec Process (Dynatec Corporation Ltd)	Mineral dissolution occurs under O_2 pressure at 100°C and 150°C, respectively for Cu_2S and $CuFeS_2$.	/	/	Originally designed to treat zinc concentrates and complex gold ores. Solid leached residue is either recycled or floated, followed by further dissolution for high Cu recovery.

(*Continued*)

Table 6.2 Summary of some of the commercial processes. (*Continued*)

Lixiviant type (Sulfate or chloride)	Name of the commercial process and country	Specific characteristics (temp, Pres…)	Cu recovery	Equations	Observations
Sulfate	MIM Albion Process (MIM and Highlands JV)	Ultrafine-ground concentrate undergoes ferric sulfate leach (~80°C) at atmospheric pressure in stirred tank.	>98% Cu recovery for chalcopyrite indicated.	/	Pilot plant run in Brisbane for Zn production.

(*Continued*)

Table 6.2 Summary of some of the commercial processes. (*Continued*)

Lixiviant type (Sulfate or chloride)	**Name of the commercial process and country**	**Specific characteristics (temp, Pres...)**	**Cu recovery**	**Equations**	**Observations**
Chloride	Intec Process (Intec Ltd)	Leach circuit operates at 85°C with oxygen supplied via air. Three-stage leach circuit of fine-ground concentrate, Halex-laden lixiviant ($BrCl_2^-$) with purification via conversion of cupric ions to cuprous, silver amalgam removal, and Fe removal via pH control.	/	/	Copper produced as dendrites, requiring remelting to produce LME ingots. Successful pilot plant demonstrations in Sydney and Tasmania.

6.8.2 Development in the Field of Copper Sulfide Mineral Leaching

In order to overcome the passivation phenomenon of the $CuFeS_2$ mineral, a number of different leaching processes have emerged and have been commercialized. They can be subdivided into two main groups, namely sulfate and chloride, and have revealed successful Cu dissolution from $CuFeS_2$ concentrate. The sulfate group can be categorized further based on the working pressure being either atmospheric or super atmospheric [78]. Table 6.2 shows the list of some of the commercial processes.

6.9 Conclusion

It is common practice for many mining companies to extract copper from chalcopyrite through the pyrometallurgical process, which consumes a lot of energy, therefore making the process relatively costly. However, the hydrometallurgical process, which could be relatively cheaper is so far less effective because of the formation of passivation layers that form around the chalcopyrite matrix preventive the dissolution of copper. The above review has covered a number of compositional, structural, and mineralogical changes that result to passivation including the formation of thick elemental sulfur layer, the precipitation of iron and formation of transition species, the formation of Fe-deficient polysulfide, and the presence gangue-related minerals belonging to carbonates, which can negatively affect the copper leaching kinetics.

It is therefore important for future studies to investigate the electrochemical behavior of chalcopyrite leaching mixture to develop conditions that can prevent the formation of passivation during the leaching process, and therefore, enhance the mobilization of copper from the chalcopyrite surface.

References

1. Watling, H.R., Hydrometallurgy chalcopyrite hydrometallurgy at atmospheric pressure: 1. Review of acidic sulfate, sulfate–chloride and sulfate–nitrate process options. *Hydrometallurgy*, 140, 163–180, 2013.
2. Tshilombo, A.F., *Mechanism and Kinetics of Chalcopyrite Passivation and Depassivation During Ferric and Microbial Leaching*, British Columbia, 2004, Available at: https://open.library.ubc.ca/collections/831/items/1.0078957.

3. Dutrizac, J.E., Ferric ion leaching of chalcopyrites from different localities. *J. Electron. Mater.*, 20, 12, 303–309, 1991.
4. Córdoba, E.M., Muñoz, J.A., Blázquez, M.L., González, F., Ballester, A., Leaching of chalcopyrite with ferric ion. Part I: General aspects. *Hydrometallurgy*, 93, 3–4, 81–87, 2008.
5. Fu, K.B., Lin, H., Mo, X.L., Dong, Y.B., Zhou, L., Study on bioleaching of different types of chalcopyrites. *J. Univ. Sci. Technol. Beijing*, 33, 7, 806–811, 2011.
6. Li, Y., Kawashima, N., Li, J., Chandra, A.P., Gerson, A.R., A review of the structure, and fundamental mechanisms and kinetics of the leaching of chalcopyrite. *Adv. Colloid Interface Sci.*, 197–198, 2013.
7. Olvera, O.G., Rebolledo, M., Asselin, E., Hydrometallurgy atmospheric ferric sulfate leaching of chalcopyrite: Thermodynamics, kinetics and electrochemistry. *Hydrometallurgy, Elsevier B.V.*, 165, 148–158, 2016.
8. Watling, H.R., Chalcopyrite hydrometallurgy at atmospheric pressure: 1. Review of acidic sulfate, sulfate-chloride and sulfate-nitrate process options. *Hydrometallurgy*, 140, 163–180, 2013.
9. Ruiz-Sánchez, Á. and Lapidus, G.T., Hydrometallurgy study of chalcopyrite leaching from a copper concentrate with hydrogen peroxide in aqueous ethylene glycol media. *Hydrometallurgy*, 169, 192–200, 2017.
10. Munoz, P.B., Miller, J.D., Wadsworth, M.E., Reaction mechanism for the acid ferric sulfate leaching of chalcopyrite. *Metallurgical Trans. B*, 10, 2, 149–158, 1979.
11. Sokić, M.D., Marković, B., Živković, D., Kinetics of chalcopyrite leaching by sodium nitrate in sulphuric acid. *Hydrometallurgy*, 95, 3–4, 273–279, 2009.
12. Burdick, C. and Ellis, J.H., The crystal structure of chalcopyrite determined by X-rays, in: *Ternary Chalcopyrite Semiconductors: Growth, Electronic Properties, and applications*, pp. 3–74, 1917.
13. Hall, S.R. and Stewart, J.M., The crystal structure refinement of chalcopyrite, CuFeS2. *Acta Crystallogr.*, 29, 3, 579–585, 1973.
14. Naderi, H., Abdollahy, M., Mostouf, N., Koleini, M.J., Shojaosadati, S.A., Manafi, M., Kinetics of chemical leaching of chalcopyrite from low grade copper ore: Behavior of different size fractions. *Int. J. Miner. Metall. Mater.*, 18, 6, 638–645, 2011.
15. Harmer-Bassell, S.L., Thomas, J.E., Fornasiero, D., Gerson, A.R., The evolution of surface layers formed during chalcopyrite leaching. *Geochim. Cosmochim. Acta*, 70, 17, 4392–4402, 2006.
16. Li, Y., Kawashima, N., Li, J., Chandra, A.P., Gerson, A.R., A review of the structure, and fundamental mechanisms and kinetics of the leaching of chalcopyrite. *Adv. Colloid Interface Sci.*, 197–198, 2013.
17. Klauber, C., Parker, A., Van Bronswijk, W., Watling, H., Sulphur speciation of leached chalcopyrite surfaces as determined by x-ray photoelectron spectroscopy. *Int. J. Miner. Process.*, 62, 1–4, 65–94, 2001.

18. Salehi, S., Noaparast, M., Ziaeddin, S., The role of catalyst in chalcopyrite passivation during leaching. *Int. J. Min. Geo-Eng.*, 51, 1, 91–96, 2017.
19. Ghahremaninezhad, A., *The Surface Chemistry of Chalcopyrite During Electrochemical Dissolution*, PhD Thesis, University of British Columbia, 2012.
20. Debernardi, G. and Carlesi, C., Chemical-electrochemical approaches to the study passivation of chalcopyrite. *Miner. Process. Extr. Metall. Rev.*, 34, 1, 10–41, 2013.
21. Zhao, H., Zhang, Y., Zhang, X., Qian, L., Sun, M., Yang, Y., Zhang, Y., Wang, J., Kim, Y., Qiu, G., The dissolution and passivation mechanism of chalcopyrite in bioleaching: An overview. *Miner. Eng.*, 136, 932, 140–154, 2019.
22. Tkácová, K. and Baláž, P., Structural and temperature sensitivity of leaching of chalcopyrite with iron(III) sulfate. *Hydrometallurgy*, 21, 1, 103–112, 1988.
23. Klauber, C., A critical review of the surface chemistry of acidic ferric sulfate dissolution of chalcopyrite with regards to hindered dissolution. *Int. J. Miner. Process.*, 86, 1–17, 2008.
24. Nazari, G. and Asselin, E., Hydrometallurgy morphology of chalcopyrite leaching in acidic ferric sulfate media. *Hydrometallurgy*, 96, 3, 183–188, 2009.
25. Hu, K.T., Gu, G.H., Li, S.K., Qiu, G.Z., Bioleaching of chalcopyrite by Leptospirillum ferriphilum. *J. Cent. South Univ.*, 19, 6, 1718–1723, 2012.
26. Zhao, H., Hu, M.H., Li, Y., Zhu, S., Qin, W., Qui, G., Wang, J., Electrochemical dissolution of chalcopyrite concentrates in stirred reactor in the presence of acidithiobacillus ferrooxidans. *Int. J. Electrochem. Sci.*, 10, 1, 848–858, 2015.
27. Petrovic, S.J., Bogdanovic, G.D., Antonijevic, M.M., Leaching of chalcopyrite with hydrogen peroxide in hydrochloric acid solution. *Trans. Nonferrous Met. Soc China (English Edition)*, 28, 7, 1444–1455, 2018.
28. Linge, H.G., A study of chalcopyrite dissolution in acidic ferric nitrate by potentiometric titration. *Hydrometallurgy*, 2, 1, 51–64, 1976.
29. Parker, A.J., Paul, R.L., Power, G.P., Electrochemistry of the oxidative leaching of copper from chalcopyrite. *J. Electroanal. Chem.*, 118, C, 305–316, 1981.
30. Joe, S., Chida, T., Sakoda, M., Nakamura, H., Tamura, M., Sato, N., Effect of sulfuric acid concentration on chalcopyrite concentrate chemical leaching. *Adv. Mater. Res.*, 71–73, 2009.
31. Acero, P., Cama, J., Ayora, C., Asta, M.P., Chalcopyrite dissolution rate law from pH 1 to 3. *Geol. Acta*, 7, 3, 389–397, 2009.
32. Córdoba, E.M., Muñoz, J.A., Blázquez, M.L., González, F., Ballester, A., Passivation of chalcopyrite during its chemical leaching with ferric ion at 68°C. *Miner. Eng.*, 22, 3, 229–235, 2009.
33. Vilcáez, J., Suto, K., Inoue, C., Bioleaching of chalcopyrite with thermophiles: Temperature–pH–ORP dependence. *Int. J. Miner. Process.*, 88, 1–2, 37–44, 2008.

34. Stott, M.B., Watling, H.R., Franzmann, P.D., Sutton, D., Role of iron-hydroxy precipitates in the passivation of chalcopyrite during bioleaching. *Minerals Eng.*, 13, 10, 1117–1127, 2000.
35. Hackl, R.P., Dreisinger, D.B., Peters, L., King, J.A., Passivation of chalcopyrite during oxidative leaching in sulfate media. *Hydrometallurgy*, 39, 1–3, 25–48, 1995.
36. Kimball, B.E., Rimstidt, J.D., Brantley, S.L., Chalcopyrite dissolution rate laws. *Appl. Geochem.*, 25, 7, 972–983, 2010.
37. Tshilombo, O.M. and Ojumu, T.V., Investigation of the effect of pH operating conditions on bioleaching of low-grade chalcopyrite in column reactors. *Adv. Mater. Res.*, 825, 401–405, 2013.
38. Butler, R.D., Carbonate leaching of uranium ores: A review. *Aaec Symposium on Uranium Processing*, February 1958, 1970, Available at: http://www.iaea.org/inis/collection/NCLCollectionStore/_Public/04/050/4050494.pdf.
39. Free, M.L., Understanding acid consumption and its relationship with copper recovery. *Proceedings of the SME Annual Meeting*, Phoenix, AZ, USA, vol. 28, pp. 74–77, 2010.
40. Casas, J.M., Crisóstomo, G., Cifuentes, L., Speciation of the Fe(II)–Fe(III)–H2 SO4–H2O system at 25 and 50 °C. *Hydrometallurgy*, 80, 4, 254–264, 2005.
41. Yue, G. and Asselin, E., Kinetics of ferric ion reduction on chalcopyrite and its influence on leaching up to 150 °C. *Electrochim. Acta*, 146, 307–321, 2014.
42. Yue, G., *Speciation of The Sulfuric Acid-Ferric Sulfate-Ferrous Sulfate-Water System And Its Application To Chalcopyrite Leaching Kinetics Up to 150 °C*, Doctoral Dissertation, Univversity of British Columbia, 2015.
43. Hirato, T., Majima, H., Awakura, Y., The leaching of chalcopyrite with ferric sulfate. *Metallurgical Trans. B*, 18, 3, 489–496, 1987.
44. Sapieszko, R.S., Patel, R.C., Matijevic, E., Ferric hydrous oxide sols. 2. Thermodynamics of aqueous hydroxo and sulfato ferric complexes. *J. Phys. Chem.*, 81, 11, 1061–1068, 1977.
45. Hiroyoshi, N., Miki, H., Hirajima, T., Tsunekawa, M., Enhancement of chalcopyrite leaching by ferrous ions in acidic ferric sulfate solutions. *Hydrometallurgy*, 60, 3, 185–197, 2001.
46. Antonijevic, M.M. and Bogdanovic, G.D., Investigation of the leaching of chalcopyritic ore in acidic solutions. *Hydrometallury*, 73, 3–4, 245–256, 2004.
47. Gómez, C., Figueroa, M., Muñoz, J., Blázquez, M.L., Electrochemistry of chalcopyrite. *Hydrometallurgy*, 43, 1–3, 331–344, 1996.
48. Dutrizac, J.E., The dissolution of chalcopyrite in ferric sulfate and ferric chloride media. *Metallurgical Trans. B*, 12, 2, 371–378, 1981.
49. Havlik, T. and Skrobian, M., Acid leaching of chalcopyrite in the presence of ozone. *Can. Metall. Q.*, 29, 2, 133–139, 1990.
50. Salehi, S., Noaparast, M., Shafaie, S.Z., Kinetics of chalcopyrite galvanic leaching using sulfate medium at low temperature in the Galvanox TM process. *Int. J. Min. Geo-Eng.*, 50, 2, 157–161, 2016.

51. Tanda, B.C., Eksteen, J.J., Oraby, E.A., O'Connor, G.M., The kinetics of chalcopyrite leaching in alkaline glycine/glycinate solutions. *Miner. Eng.*, 135, 118–128, 2019.
52. Juhász, A.Z. and Opoczky, L., *Mechanical Activation of Minerals by Grinding, Pulverizing and Morphology of Particles*, Akadémiai Kiadó, Budapest, 1990.
53. Balaz, P., Extractive metallurgy of activated minerals. *Miner. Eng.*, 13, 10–11, 1191, 2000.
54. Hirato, T., Majima, H., Awakura, Y., The leaching of chalcopyrite with cupric chloride. *Metallurgical Trans. B*, 18, 1, 31–39, 1987.
55. Córdoba, E.M., Muñoz, J.A., Blázquez, M.L., González, F., Ballester, A., Leaching of chalcopyrite with ferric ion. Part II: Effect of redox potential. *Hydrometallurgy*, 93, 3–4, 88–96, 2008.
56. Kametani, H. and Aoki, A., Effect of suspension potential on the oxidation rate of copper concentrate in a sulfuric acid solution. *Metallurgical Trans. B*, 16, 4, 695–705, 1985.
57. Koleini, S.M.J., Aghazadeh, V., Sandström, Å., Acidic sulfate leaching of chalcopyrite concentrates in presence of pyrite. *Miner. Eng.*, 24, 5, 381–386, 2011.
58. Sandstrom, A., Shchukarev, A., Paul, J., XPS characterisation of chalcopyrite chemically and bio-leached at high and low redox potential. *Miner. Eng.*, 18, 505–515, 2005.
59. Santos, A.L.A., Arena, F.A., Benedetti, A.V., Bevilaqua, D., Effect of redox potential on chalcopyrite dissolution imposed by addition of ferrous ions. *Eclet. Quim.*, 42, 1, 40–50, 2017.
60. Southwood, M. and Cairncross, B., The minerals of Palabora, Limpopo Province, South Africa. *Rocks Miner.*, 92, 5, 426–452, 2017.
61. Sracek, O., Mihaljevic, M., Kríbek, B., Majer, V., Veselovsky, F., Geochemistry and mineralogy of Cu and Co in mine tailings at the Copperbelt, Zambia. *J. Afr. Earth Sci.*, 57, 1–2, 14–30, 2010.
62. Miller, J.D., McDonough, P.J., Portillo, H.Q., Electrochemistry in silver catalyzed ferric sulfate leaching of chalcopyrite, in: *Process and Fundamental Considerations of Selected Hydrometallurgical Systems*, vol. 27, p. 327, SME-AIME, New York, 1981.
63. Lu, Z.Y., Jeffrey, M., II, Lawson, F., An electrochemical study of the effect of chloride ions on the dissolution of chalcopyrite in acidic solutions. *Hydrometallurgy*, 56, 2, 145–155, 2000.
64. Nicol, M., Miki, H., Velásquez-Yévenes, L., The dissolution of chalcopyrite in chloride solutions: Part 3. Mechanisms. *Hydrometallurgy*, 103, 1–4, 86–95, 2010.
65. Liu, Q., Chen, M., Yang, Y., Electrochimica Acta The effect of chloride ions on the electrochemical dissolution of chalcopyrite in sulfuric acid solutions. *Electrochim. Acta*, 253, 257–267, 2017.
66. Carneiro, M.F.C. and Leão, V.A., The role of sodium chloride on surface properties of chalcopyrite leached with ferric sulfate. *Hydrometallurgy*, 87, 3–4, 73–82, 2007.

67. Arce, E.M. and González, I., A comparative study of electrochemical behavior of chalcopyrite, chalcocite and bornite in sulfuric acid solution. *Int. J. Miner. Process.*, 67, 1–4, 17–28, 2002.
68. Warren, G.W. and Wadsworth, M.E., Passive and transpassive anodic behavior of chalcopyrite in acid solutions. *Metallurgical Trans. B*, 13, 571–579, 1982.
69. Liang, C.L., Xia, J.L., Nie, Z.Y., Yu, S.J., Xu, B.Q., Effect of initial pH on chalcopyrite oxidation dissolution in the presence of extreme thermophile Acidianus manzaensis. *Trans. Nonferrous Met. Soc China (English Edition)*, 24, 6, 1890–1897, 2014.
70. Majuste, D., Ciminelli, V.S.T., Eng, P.J., Osseo-Asare, K., Applications of in situ synchrotron XRD in hydrometallurgy: Literature review and investigation of chalcopyrite dissolution. *Hydrometallurgy*, 131–132, 2013.
71. Muszer, A., Wódka, J., Chmielewski, T., Matuska, S., Covellinisation of copper sulfide minerals under pressure leaching conditions. *Hydrometallurgy*, 137, 1–7, 2013.
72. Saavedra, A., García-Meza, J.V., Corton, E., Gonzalez, I., Understanding galvanic interactions between chalcopyrite and magnetite in acid medium to improve copper (Bio) leaching. *Electrochim. Acta*, 265, 569–576, 2018.
73. Holmes, P.R. and Crundwell, F.K., Kinetic aspects of galvanic interactions between minerals during dissolution. *Hydrometallurgy*, 39, 1–3, 353–375, 1995.
74. Dixon, D.G., Mayne, D.D., Baxter, K.G., Galvanox™-a novel galvanically-assisted atmospheric leaching technology for copper concentrates. *Can. Metall. Q.*, 47, 3, 327–336, 2008.
75. Koleini, S.M.J., Jafarian, M., Abdollahy, M., Aghazadeh, V., Galvanic leaching of chalcopyrite in atmospheric pressure and sulfate media: Kinetic and surface studies. *Ind. Eng. Chem. Res.*, 49, 13, 5997–6002, 2010.
76. Majuste, D., Ciminelli, V.S.T., Osseo-Asare, K., Dantas, M.S.S., Hydrometallurgy quantitative assessment of the effect of pyrite inclusions on chalcopyrite electrochemistry under oxidizing conditions. *Hydrometallurgy*, 113–114, 2012.
77. Nyembwe, K.J., Fosso-Kankeu, E., Waanders, F., Nyembwe, K.D., Structural, compositional and mineralogical characterization of carbonatitic copper sulfide: Run of mine, concentrate and tailings. *Int. J. Miner. Metall. Mater.*, 26, 2, 143–151, 2019.
78. Dreisinger, D., Copper leaching from primary sulfides: Options for biological and chemical extraction of copper. *Hydrometallurgy*, 83, 1–4, 10–20, 2006.

7

Fe^{3+}-Fe^{2+} Redox Cycle Peculiarity in the Acid Dissolution of Copper–Cobalt Complex Ores

Bienvenu Mbuya[1,2,3], Léon Zeka[2,4,5] and Antoine F. Mulaba-Bafubiandi[3,6*]

[1]*Department of Metallurgy, Faculty of Engineering, University of Likasi, Likasi, Haut-Katanga, Democratic Republic of Congo*
[2]*Gécamines Metallurgical Research Centre, Likasi, Haut-Katanga, Democratic Republic of Congo*
[3]*Mineral Processing and Technology Research Centre, Department of Metallurgy, School of Mining, Metallurgy and Chemical Engineering, Faculty of Engineering and The Built Environment, University of Johannesburg, Johannesburg, South Africa*
[4]*Department of Chemistry, Faculty of Engineering, University of Lubumbashi, Lubumbashi, Haut-Katanga, Democratic Republic of Congo*
[5]*Centre de Recherche et d'Expertise ZEBRE – CreZ, Kolwezi, Lwalaba, Democratic Republic of Congo*
[6]*Faculte des Sciences Appliquees, Universite de Mbuji-Mayi, Mbuji-Mayi, Kasai, Republique Democratique du Congo*

Abstract
Cobalt, naturally found in association with copper, as it is the case in the copper–cobalt bearing minerals from the African Copperbelt region, as found in the DRC-Zambia chain, has been often considered as a by-product from copper extraction. With Co(III) in the ore, the hydrometallurgical processing of such an ore goes through the use of reducing agents as metabisulfite and sulfur dioxide at a time preceded by a microwave energy pretreatment. The abovementioned processing leads to the inevitable release of sulfur dioxide into the environment with the expected negative impacts. This chapter discusses a polluting reagent free copper–cobalt complex oxide and sulfide ores acid green dissolution process.

**Corresponding author*: antoinemulaba1@gmail.com

Elvis Fosso-Kankeu, Bhekie B. Mamba and Antoine F. Mulaba-Bafubiandi (eds.) Recovery of Values from Low-Grade and Complex Minerals: Development of Sustainable Processes, (179–216) © 2024 Scrivener Publishing LLC

The use of the polluting reducing agents has been substituted by leveraging on the Fe^{3+}-Fe^{2+} *in-situ* redox cycle. Starting with an overview of conventional dissolution processes, the chapter emphasizes on the sound thermodynamics study of the 'mineral–mineral' process to discuss the development of this Co(III) eco-friendly dissolution from complex copper–cobalt ores, before ending with typical examples.

Keywords: Green leaching process, redox cycle Fe^{3+}-Fe^{2+}, mineral–mineral process, copper–cobalt complex ores, heterogenite, Copperbelt DRC-Zambia

7.1 Introduction

Cobalt is commonly listed among strategic metals. Currently, the demand for cobalt is rapidly growing, following its use in green technologies, particularly in the cathodes of lithium-ion batteries and nickel metal hydride batteries used in electric or hybrid vehicles. Cobalt is also used as a component of superalloys used in gas turbines and nuclear reactors, as well as in wind turbine magnets, radar magnets, missile guidance systems, marine propulsion systems, and sensors.

In nature, cobalt is found in oxide and sulfide ores. The large reserve estimated at one-third of the world's reserves is found in the DRC-Zambia African Copperbelt region. In this region, after copper, it is the cobalt found in the copper–cobalt ores that is the main metal valued. After hydrometallurgical or pyrometallurgical treatment, cobalt derivatives are produced in various forms including cobalt hydroxides ($Co(OH)_2$), cobalt sulfate ($CoSO_4$), cobalt carbonate ($CoCO_3$), cathode cobalt (Co^{o}), or white alloy (Cu^0-Co^0). However, due to the increased demand for this metal in energy storage technologies, the $Co(OH)_2$ form is the most required.

For the hydrometallurgical extraction of cobalt from oxidized ore, the first step includes acid dissolution through leaching, as shown in reactions 7.1 to 7.3. The pregnant leaching solutions are first purified, then followed by the copper and cobalt precipitation as salts and/or metals, according to the following flow sheet (Figure 7.1) [1, 2].

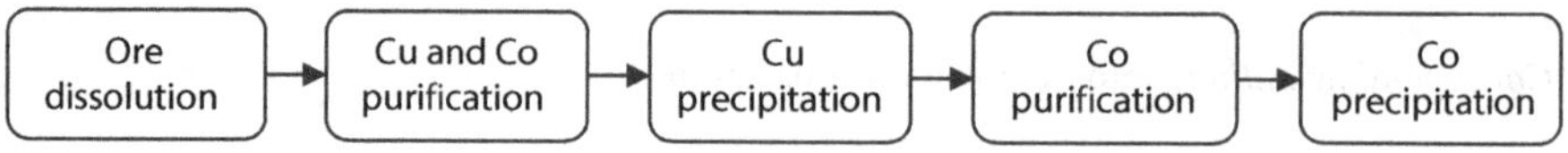

Figure 7.1 Copper and cobalt extraction process route from oxidized ore.

The Cu(II) and the Co(II) go into solution by using the acid:

$$CuCO_3.Cu(OH)_2 + 2H_2SO_4 \rightarrow 2CuSO_4 + CO_2 + 3H_2O \quad (7.1)$$

$$CoO + H_2SO_4 \rightarrow CoSO_4 + H_2O \quad (7.2)$$

But for Co(III), in addition to the acid, a reducing agent like Fe(II) is required:

$$Co_2O_3 + 2FeSO_4 + 3H_2SO_4 \rightarrow 2CoSO_4 + Fe_2(SO_4)_3 + 3H_2O \quad (7.3)$$

As Co(III), cobalt is found in heterogenite [3]. In practice, the most commonly used reducing reagents are copper in the metallic state, metabisulfite ($Na_2S_2O_5$), sulfur dioxide, ferrous iron, ions added in the form of $FeSO_4$ and iron filings. Each reducing reagent presents some advantages and disadvantages [4].

Mwema *et al.* [5] studied the Co(III) dissolution in the presence of $Na_2S_2O_5$ as a reducing agent. The metallurgical performances were satisfactory, although the environmental impact related to the SO_2 emission requires the development of other dissolution alternatives. In this perspective, the use of Fe^{2+} instead of $Na_2S_2O_5$ was investigated abundantly. Zeka [3] studied heterogenite leaching using Fe^{2+} from pyrite (FeS_2) that was added to the reaction medium. These Fe^{2+} ions oxidize as Co(III) becomes reduced. The Fe^{3+} ions are reduced to Fe^{2+} ions due to the presence of pyrite, thus creating an Fe^{3+}-Fe^{2+} loop, which ultimately favors the leaching of Co(III). Mbuya *et al.* [6] exploited the same Fe^{3+}-Fe^{2+} loop by creating it in the reaction medium from a mixed ore containing both oxide minerals and sulfide minerals. These studies showed that it was possible to dissolve Co(III) using Fe^{2+} ions from an Fe-bearing mineral. Finally, Mbuya *et al.* [7] compared the dissolution of Co in the presence of different reducing agents like Cu, $Na_2S_2O_5$, $FeSO_4$, and Fe. In literature, significant advances have been made in the modeling of cobalt leaching. To this end, several approaches and techniques were used. These include purely mathematical approaches such as the response surface methodology [9, 11], statistical approaches, such as that of Taguchi [1], principal component analysis [7, 8], conditional probabilistic approaches [11], or artificial intelligence approaches such as artificial neural networks [9, 12].

Processes leading to the SO_2 emission are not environment-friendly. All other processes should be encouraged to both protect nature and optimally dissolve cobalt in order to meet the great global demand for this metal. In this chapter, all these processes are discussed. A peculiar emphasis is placed on the processes in which the Fe^{3+}-Fe^{2+} loop is upgraded and maintained in order to leach the complex ores carrying Co(III) in a process called "mineral–mineral".

This chapter is devoted to the theory on the conventional methods of leaching of copper–cobalt ores before proving the existence of the Fe^{3+}-Fe^{2+} loop in some reaction systems leading to a "**mineral–mineral process**". Finally, some practical examples of this process are given at the end of the chapter.

7.2 Conventional Leaching of Copper–Cobalt Minerals

Conventional methods and techniques for leaching copper–cobalt minerals are discussed here. These methods strongly depend on the nature of the minerals present, and are based on the use of Pourbaix diagrams, which constitute an essential tool in understanding reactions in hydrometallurgy. To date, the solution reactions of copper and cobalt are well understood in the mining industry. Leaching yields in a stirred tank reach values above 95% for these two metals. However, the mineralogical variability that is responsible for the dispersion of the results associated with environmental aspects creates an environment in which research on leaching of copper–cobalt minerals remains permanent. The carrier minerals of Cu and Co are numerous with varied mineralogical forms. They are leached using several conventional techniques.

7.2.1 Minerals Found in Copper–Cobalt Ores

A large Cu and Co reserve is found in the DRC-Zambia Copperbelt. In this region of the globe, the minerals carrying these two metals are contained in the sedimentary rocks of Neoproterozoic age of the Katanguian Supergroup. Over time, significant advances have been made in understanding the mineralization of the copper arc deposits as explained below.

7.2.1.1 Location of the African Copperbelt

Situated between the southeastern part in the Democratic Republic of Congo and northern Zambia, the Central African Copperbelt contains

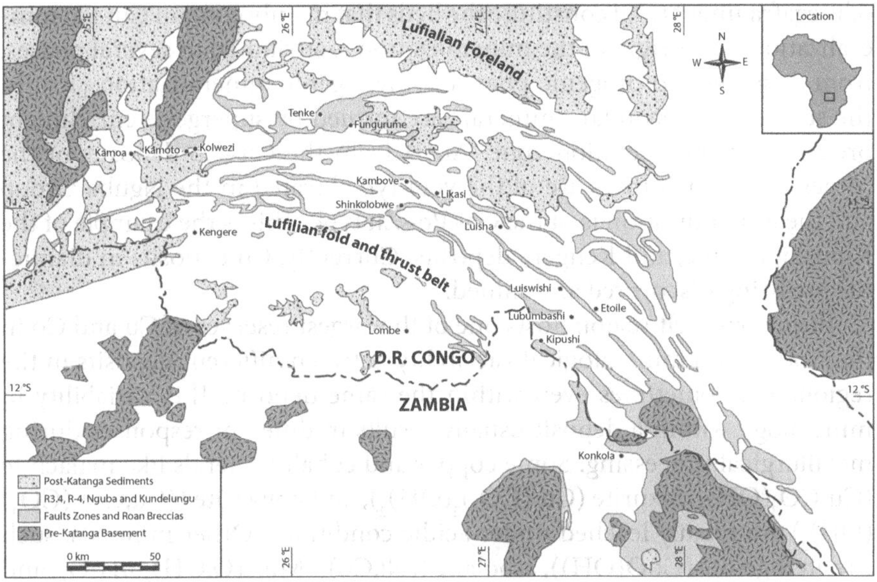

Figure 7.2 Location of Cu–Co deposits along the Lufilian arc [13].

world-class Cu-Co and Cu-Pb-Zn deposits. On the side of the Democratic Republic of Congo, the Copperbelt extends over approximately 500 km long and 50 km wide. Several deposits with complex structures, different sizes, and depths are present along faults and folds deformed during the Lufilian orogeny (Figure 7.2).

7.2.1.2 Geology of the Katanguian

The Katanguian Supergroup is subdivided into four groups, from bottom to top: the Roan (R), Nguba (Ng), Kundelungu (Ku), and Biano groups. This supergroup is characterized by a succession of carbonate and detrital sedimentary rocks with a thickness of about 15 km. The Roan is made up of platform-type carbonate deposits with detrital bed intercalations. It is subdivided from bottom to top by the Musonoi, Mines, Fungurume, and Mwashya Subgroup. The Nguba Group is separated from the Roan Group by the Mwale Formation, while the Kyandamu Formation is located at the base of the Kundelungu Group. The newly defined Biano Group is characterized by a succession of sandstone, shale, and conglomerate. Traces of Cu–Co mineralization are found in the Dipeta subgroup. The Mining Subgroup is mainly composed of more or less argillaceous dolomites with horizons of shale, sandstone, and conglomerate. They outcrop in the form of megafragments and constitute the beds that mainly contain Cu–Co mineralization. The origin of the mineralization is attributed to several phases, from diagenesis to orogenic phases by hypogenic mineralization [13]. In the subsurface, secondary minerals are formed by supergene enrichment processes in the alteration zone. Apart from the Cu–Co deposits of the Mines subgroup, others are of the vein type located in the Nguba Group and the mineralization is of the Cu–Pb–Zn type. This is the example of the Kipushi, Lombe, and Kengere deposits. Currently, Cu deposits in diamictite are being discovered and mined.

The Copperbelt region hosts one of the largest reserves of Cu and Co in the world, with mineralogical variability between different deposits in the region, and sometimes even within the same deposit. This variability in mineralogy within a deposit usually results in different responses during metallurgical processing. Some copper and cobalt minerals like malachite ($Cu_2CO_3(OH)_2$), azurite ($Cu_3(CO_3)_2(OH)_2$), and kolwezite ($(Cu,Co)_2(CO_3)(OH)_2$) are readily leached under acidic conditions. Other minerals, such as heterogenite ($CoO(OH)$), asbolane ($(Ni,Co)_{2-x}Mn^{4+}(O,OH)_4.nH_2O$), and chrysocolla ($Cu_2H_2Si_2O_5(OH)_4$), require rather specific conditions with the presence of reducing agents or strong acids for their leaching [1, 2, 14]. Leaching of Cu and Co becomes even more difficult when they are locked in refractory gangue minerals, such as clays and micas. Furthermore, the mineralogy of sulfide ores is also very complex. The various mining centers and quarries contain a great diversity of mineral species. In the Congolese Copperbelt, carrollite ($CuCo_2S_4$), siegenite ($(Ni,Co)_3S_4$), and linnaeite ($Co^{2+}Co^{3+}{}_2S_4$) are the most abundant primary cobalt ore minerals. Chalcopyrite ($CuFeS_2$), bornite (Cu_5FeS_4), chalcosine (Cu_2S), and covellite (CuS) are the main copper minerals. Sphalerite (ZnS) is the main mineral for zinc. There are also abundant quantities of galena (PbS) and pyrite (FeS_2). As with heterogenite, chalcopyrite and carrollite are particularly refractory; even in the presence of an oxidant, they are difficult to leach. Many studies have focused on the leaching of chalcopyrite and other copper sulfides, in general. Regarding carrollite and cobalt sulfides, very few studies have been devoted to them [15].

7.2.2 Thermodynamics of the Cu and Co Bearing Mineral Dissolution

The leaching equilibria of Cu and Co from their solid form in the ores can be easily visualized on the Pourbaix diagrams. Only species involved in

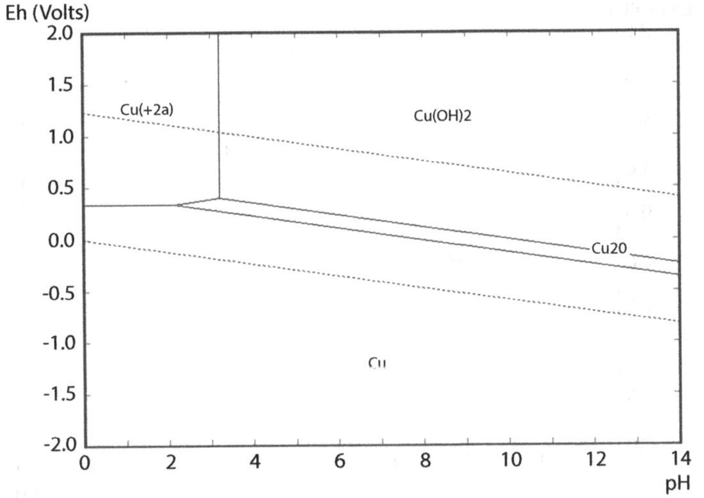

Figure 7.3 Eh–pH diagram of copper–water system at 25°C (HSC Chemistry 6.0).

important reactions are given on the diagrams below to understand the behavior of Cu and Co in hydrometallurgy.

7.2.2.1 Potential–pH Diagram of the Cu–H_2O System at 25°C

The Pourbaix diagram of the Cu–H_2O system at 25°C is given in Figure 7.3. In solution, copper is in the form of Cu^{2+}.

In the solid state in ores, copper is found in the simplified form of $Cu(OH)_2$ at the +II stage and Cu_2O at the +I stage. This diagram shows that Cu^{2+} ions are stable at low pH. To leach, it is therefore necessary to work in an acid medium at pH values corresponding to the stability zone of the Cu^{2+} ions.

7.2.2.2 Potential–pH Diagram of the Co–H_2O System at 25°C

The Pourbaix diagram of the Co–H_2O system at 25°C is given in Figure 7.4. The Co^{2+} ion is stable in an acid medium at pH values below 7, while the Co^{3+} ion only exists in highly oxidizing and very acidic conditions. In an aqueous medium, only the Co^{2+} ion exists. Under these conditions, to solubilize the simplified $Co(OH)_2$ form of cobalt at the +II stage, it is just necessary to add acid as in the previous case of copper. But to solubilize

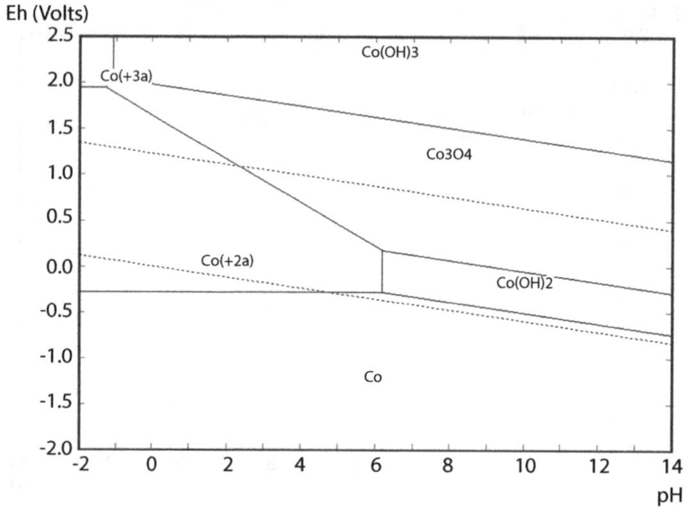

Figure 7.4 Eh–pH diagram of cobalt–water system at 25°C (HSC Chemistry 6.0).

the simplified Co_2O_3 form of cobalt at the +III stage, the operation must be carried out under reducing conditions in the presence of a reducing agent.

7.2.2.3 Leaching Reactions

Leaching is a process of selective dissolution, in ionic form, of the metals sought from mineral materials. To put the metals in solution, it is necessary to use an appropriate solvent called a leaching agent. Its selection is dictated by several factors including selectivity or minimal solubilization of impurities:

- the nature of the minerals present;
- its corrosive power in relation to leaching reactors or equipment maintenance cost;
- its cost and availability on the market;
- its specific consumption; and
- the possibility of its recovery or even of its regeneration, etc.

The leaching process involves the following steps:

- the diffusion of the leaching agent from the solution to the surface of the ore, where it is adsorbed;

- the chemical reaction taking place on the surface of the ore leading to the formation of a soluble species; and
- the desorption of the reaction products from the surface and their diffusion towards the solution.

Different leaching agents can be employed. The most common are water, acids, basics, complex solutions, reducing solutions, oxidizing solutions, and bacterial solutions. Water leaching or neutral leaching dissolves some water-soluble minerals and sulfates ($CuSO_4$) formed after roasting operations:

$$CuSO_4 + nH_2O \rightarrow Cu^{2+} + SO_4^{2-} + nH_2O \tag{7.4}$$

Water has the advantage of being inexpensive and non-corrosive. However, it can only leach a few minerals. For other cases, the use of a leaching reagent is necessary. The leaching of Cu–Co ores from the Copperbelt is well documented [1, 2, 4, 7, 12, 14, 16].

Acid leaching, with its two variants, reducing and oxidizing, is the most applied on copper–cobalt minerals. Sulfuric acid is the most used in plants based in the Katanga mining area due to its price and ability to regenerate to solvent extraction and electrolysis extraction. Depending on the nature of the minerals, the following types of leaching are applied:

- acid leaching for the oxide minerals of copper and cobalt on the +II stage;
- oxidative acid leaching for oxidized copper minerals at the +I level;
- reducing acid leaching for oxidized cobalt minerals at stage +III; and
- oxidative acid leaching for copper and cobalt sulfide minerals.

7.2.3 Leaching of Oxidized Copper Minerals

Based on the Pourbaix diagram in Figure 7.1, to leach the copper in its +II stage, acid must be added. The solubilization reactions are summarized as follows considering CuO as the simplified form:

$$CuO + 2H^{+} \rightarrow Cu^{2+} + H_2O \ \Delta G^{\circ}_{r,298} = -10.027\ Cal/mol \quad (7.5)$$

$$\log[Cu^{2+}] = 7.89 - 2pH \quad (7.6)$$

Considering the $Cu(OH)_2$ as a simplified form of copper oxidized minerals, the equation becomes as follows:

$$Cu(OH)_2 + 2H^{+} \rightarrow Cu^{2+} + 2H_2O \quad (7.7)$$

$$\log[Cu^{2+}] = 9.21 - 2pH \quad (7.8)$$

In the Cu^{+} form, considering the Cu_2O is in the simplified form that the following leaching reaction is considered:

$$Cu_2O + 2H^{+} \rightarrow 2Cu^{2+} + H_2O + 2\grave{e} \quad (7.9)$$

$$E = 0.203 + 0.059pH + 0.0591\log[Cu^{2+}] \quad (7.10)$$

Reaction 7.9 is possible only in the presence of an oxidant, otherwise the reaction becomes as follows:

$$Cu_2O + 2H^{+} \rightarrow Cu^{2+} + Cu + H_2O \quad (7.11)$$

7.2.4 Leaching of Cobalt Oxidized Minerals

In the oxidized minerals of copper–cobalt minerals, cobalt is present in both bivalent and trivalent oxidation stages [3]. Based on the Pourbaix (or Eh–pH) diagram (Figure 7.4), only Co^{2+} can be present in an aqueous medium at low pH (acidic medium). The Co^{3+} ion is only stable outside the water stability zone (Figure 7.4). The Co^{3+} ion, therefore, does not exist in the aqueous phase in the dissolved state. But Co^{3+} can exist in solid form in oxidized compounds as in Co_2O_3 or $Co(OH)_3$. The dissolution of cobalt oxide ores may be accomplished in sulfuric acid media, but cobalt in trivalent state leach only in the presence of reducing agents, which is the subject of this chapter.

7.2.4.1 *Reaction Chemistry*

Considering the useful area on the Pourbaix or Eh–pH diagram as given in Figure 7.5, three phases can be distinguished:

- a phase where the cobalt is in the ionic state: Co^{2+};
- two solid phases in which the cobalt is in the bivalent and trivalent oxidation state. Pourbaix (1963) considered that the $Co(OH)_2$ and $Co(OH)_3$ forms are more stable in aqueous media than the CoO and Co_2O_3 forms. Therefore, the author adopted the $Co(OH)_2$ and $Co(OH)_3$ forms to draw the diagram.

The thermodynamic stability domain of cobalt also presents a small zone (4th phase: Co_3O_4) located between $Co(OH)_2$ and $Co(OH)_3$. This 4th phase and the "13" equilibrium are of little interest in this study.

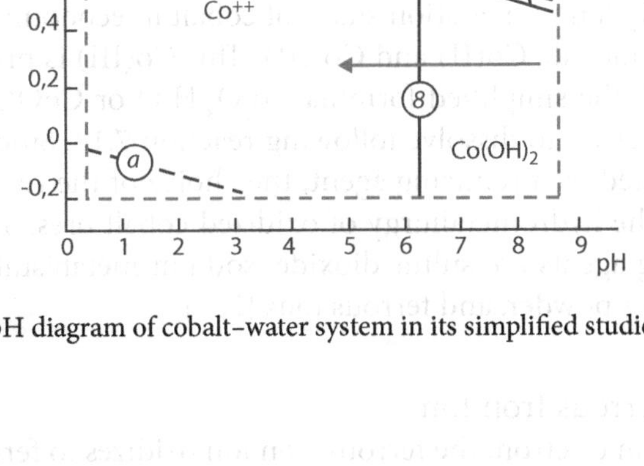

Figure 7.5 Eh–pH diagram of cobalt-water system in its simplified studied region [4].

The important balances for this study are the following:

- Equilibrium 8, between Co^{2+} and $Co(OH)_2$: only affected (or influenced) by the pH:

$$Co(OH)_{2(s)} + 2H^{+}_{(aq)} \rightarrow Co^{2+}_{(aq)} + H_2O_{(l)} \tag{7.12}$$

$$pH = \frac{1}{2}(12.6 - \log a_{Co^{2+}}) \tag{7.13}$$

- Equilibrium 15, between Co^{2+} and $Co(OH)_3$ (or Co_2O_3): influenced by the pH and Eh:

$$Co_2O_{3(s)} + 6H^{+}_{(aq)} + 2\grave{e} \rightarrow 2Co^{2+}_{(aq)} + 3H_2O_{(l)} \tag{7.14}$$

$$E = 1.746 - 0.177 * pH - 0.06 * \log a_{Co^{2+}} \tag{7.15}$$

To leach the cobalt oxides, i.e., to move the two equilibria (8 and 15) into the zone of Co^{2+} stability, as illustrated by arrows in Figure 7.5, it is necessary to decrease the pH, and therefore, add the acid. However, this condition is not sufficient to leach Co_2O_3. Its leaching is a reduction reaction requiring a reducing reagent for its realization. The performance of these reagents has been discussed by Zeka [3].

7.2.4.2 Discussions on the Reducing Agents of Co(III)

The commonly found oxidation states of cobalt in economical deposits of oxidized ores include Co(II) and Co(III). This Co(III) is present as a heterogenite with the simplified formula $Co_2O_3.H_2O$ or CoOOH. This must capture an electron to dissolve following reaction 7.14. Since the electron must be donated by a reducing agent, the choice of the reducing agent is essential for the hydrometallurgy of oxidized cobalt ores. The commonly used reducing agents are sulfur dioxide, sodium metabisulfite ($Na_2S_2O_5$), metallic copper powder, and ferrous ions [3, 7].

7.2.4.2.1 Ferrous Iron Ion

By capturing an electron, the ferrous iron ion oxidizes to ferric iron ion:

$$Fe^{3+}_{(aq)} + \grave{e} \leftrightarrow Fe^{2+}_{(aq)} \tag{7.16}$$

$$E = 0.77 + 0.06 * \log \frac{a_{Fe^{3+}}}{a_{Fe^{2+}}} \tag{7.17}$$

The Fe^{2+} ion is a strong reducing agent. In industry, iron is added in the form of Fe^0 and Fe^{2+} is obtained by dissolving iron in sulfuric acid present in the leaching reactor.

However, the use of iron is limited in hydrometallurgy because its concentration in the leachate increases because it works in a closed leaching–electrolysis extraction circuit. During electrolysis, Fe^{2+} oxidizes at the anode and Fe^{3+} is reduced at the cathode. The iron circulates in a Fe^{3+}-Fe^{2+} loop and consumes the current unnecessarily, which reduces the current efficiency.

7.2.4.2.2 Metallic Copper

Metallic copper oxidizes to Cu^{2+} :

$$Cu_{(s)} \leftrightarrow Cu^{2+}_{(aq)} + 2\grave{e} \tag{7.18}$$

$$E = 0.34 + 0.06 * \log a_{Cu^{2+}} \tag{7.19}$$

In industrial applications, copper is used in the form of powder copper (impure copper) obtained from copper electrolytic of cobalt solutions.

Although copper can reduce cobalt (III), its use is limited because the reaction is of the solid–solid type and slow. In addition, there is a risk of losing unreacted copper.

7.2.4.2.3 Sulfur Dioxide and Sodium Metabisulfite $\left(\text{Couple } SO_4^{2-}/SO_2\right)$

The sulfur dioxide gases are oxidizing in sulfate in aqueous media:

$$SO_{2(aq)} + 2H_2O \leftrightarrow SO^{2-}_{4(aq)} + 4H^{+}_{(aq)} + 2\grave{e} \tag{7.20}$$

$$E = 0.17 - 0.24 * pH + 0.06 * \log \frac{a_{SO_4^{2-}}}{p_{SO_2}} \quad (7.21)$$

When using metabisulfite, sulfur dioxide gas is released in the presence of a strong acid:

$$Na_2S_2O_{5(s)} + H_2SO_{4(aq)} \rightarrow Na_2SO_{4(aq)} + 2SO_{2(aq)} + H_2O_{(l)} \quad (7.22)$$

The reactivity of sulfur dioxide rises in the presence of Fe^{3+}, the sulfur dioxide reduces the Fe^{3+} in Fe^{2+}, then reducing cobalt (III):

$$SO_{2(aq)} + Fe_2(SO_4)_{3(aq)} + H_2O_{(l)} \rightarrow 2FeSO_{4(aq)} + 2H_2SO_{4(aq)} \quad (7.23)$$

SO_2 is widely used in metallurgy in Katanga. It is produced from metabisulfite (reaction 7.22). It is also obtained, as a by-product in the manufacture of sulfuric acid, from the roasting of elemental sulfur.

Furthermore, Ferron and Henry [17] observed a copper loss related to SO_2 additions, probably due to the precipitation of the Le Chevreult salt:

$$3CuSO_4 + SO_2 + 6H_2O \rightarrow CuSO_3.Cu_2SO_3.2H_2O + 4H_2SO_4 \quad (7.24)$$

This potential precipitation of copper sulfite represents a real possibility of loss of valuable metals. Another disadvantage of using SO_2 is the cost of obtaining it:

- when it is produced from metabisulfite that will have to be paid for [5] and
- when it is obtained from a sulfuric acid unit, it requires a cost because the establishment of said unit is an additional investment for the metallurgical industry.

The reducing agents used for cobalt (III) leaching are Fe^{2+}, Cu^0, and SO_2. Each reducing agent has its advantages and disadvantages as explained above. Taking the case of the Greater Katanga Region in the Southern-East of the DR Congo, the reducing agent most currently used is SO_2. It is produced either by metabisulfite or by the roasting of elemental sulfur.

7.2.4.3 *Environmental Aspects Related to the Use of Reagents that Generate SO_2*

The use of SO_2 has several disadvantages, the most obvious being the environmental and health risks associated with SO_2 gas emissions.

7.2.4.4 *Experimental Data on Co(III) Reduction*

Some results of Co(III) leaching are discussed here to evaluate their performance on some reducers. The study was applied to a few ore samples from DR-Congo Copperbelt ores: Kimpe (K), Luiswishi (L), Mukondo (M), and Tilwizembe (T). The following reducing agents were used in the leaching tests: Fe in the form of filings, copper in powder form, Fe^{2+} added as $FeSO_4.7H_2O$, and $Na_2S_2O_5$ in solid form [7].

7.2.4.4.1 Materials and Methods

The quantities of reducing agents necessary for the reduction of Co (III) were calculated stoichiometrically by considering Equations 7.25 to 7.28. The amounts added were 1.5, 2.0, 2.5, and 3.0 times the stoichiometric amounts for $FeSO_4$ and $Na_2S_2O_5$:

$$Co_2O_3 + Cu° + 3H_2SO_4 \rightarrow 2CoSO_4 + CuSO_4 + 3H_2O \quad (7.25)$$

$$Co_2O_3 + Fe° + 3H_2SO_4 \rightarrow 2CoSO_4 + FeSO_4 + 3H_2O \quad (7.26)$$

$$2Co_2O_3 + Na_2S_2O_5 + 3H_2SO_4 \rightarrow 4CoSO_4 + Na_2SO_4 + 3H_2O \quad (7.27)$$

$$Co_2O_3 + 2FeSO_4 + 3H_2SO_4 \rightarrow 2CoSO_4 + Fe_2(SO_4)_3 + 3H_2O \quad (7.28)$$

From the optima found, a database is created, which was defined by seven quantitative variables and 16 statistical items corresponding to 16 trials. These variables are the amount of the reducing agent, the Cu leaching yield, Co leaching yield, Cu solubilization, Co solubilization, the consumption of reducing agent by the cobalt and the redox potential at the leaching reaction end. The database thus formed was studied by the principal component analysis approach in order to derive statistical information allowing a better understanding of the leaching behavior of the ores under study. The XLSTAT® 2019 software compatible with Microsoft® Windows® 10 was used for calculations and plotting of curves.

The results of the chemical analysis for the four samples studied indicate that not all Cu and Co were in the form of oxides. Fe was also present in very large quantities. The gangue was essentially siliceous. Apart from malachite ($Cu_2CO_3(OH)_2$), the major minerals observed included chrysocola ($Cu_4H_4Si_4O_{10}(OH)_5$), heterogenite (CoO.OH), quartz (SiO_2) and dolomite ($CaCO_3.MgCO_3$). In addition, there was goethite ((FeO.OH) $3H_2O$) for the Mukondo ore, Muscovite ($KAl_2(AlSi_3O_{10})$ (OH,F)) for Luiswishi ore, and limonite ($5Fe_2O_3$ $2SiO_2.9H_2O$) for Tilwizembe ore. The chemical composition of the ores studied were Mukondo (Cu: 1.78%, Co: 1.56%, Fe: 1.66%, SiO_2: 53.95%), Kimpe (Cu: 4.55%, Co: 1.42%, Fe: 3.19%, SiO_2: 66.46%), Luiswishi (Cu: 1.23%, Co: 0.91%, Fe: 1.63%, SiO_2: 55.18%), and Tilwizembe (Cu: 1.13%, Co: 0.70%, Fe: 2.17%, SiO_2: 45.6%).

7.2.4.4.2 Findings from the Dissolution Tests

The comparison of leaching with and without the addition of a reducing agent is given in Figure 7.6. In all tests, Fe^{2+} ions were used at the stoichiometric amount. With the reducing medium, Co yield increases for all the samples. This increase in yield confirms the presence of trivalent cobalt, the Co(III) form under investigation.

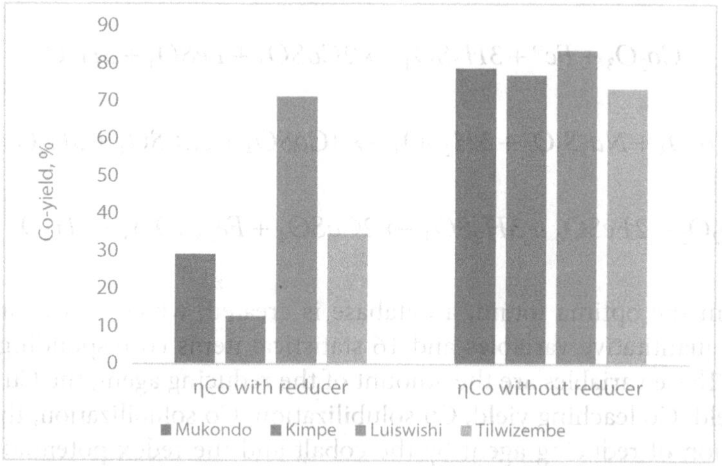

Figure 7.6 Leaching test results with and without the addition of a reducing agent [7].

Figures 7.7–7.10 show the results of the dose optimization tests of reducing agents. The optima found formed a database for the principal component analysis. This base, therefore, consisted of the optimal amounts of reducing agents, the solubilization yields of Cu and Co, the solubilization of Cu and Co, the consumption of reducing agents, and the final redox potential of all the leaching tests.

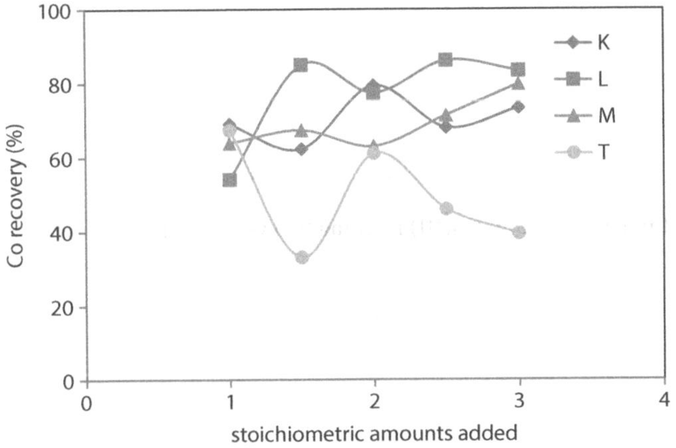

Figure 7.7 Influence of Fe^{2+} on Co(III) leaching (Co vs. Fe^{2+}) [7].

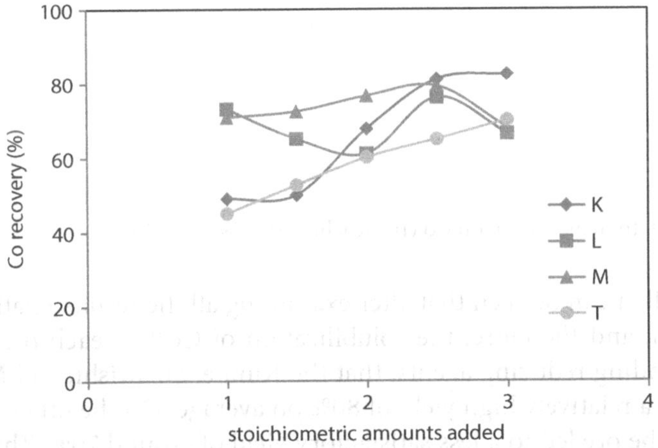

Figure 7.8 Influence of $Na_2S_2O_5$ on Co (III) leaching (Co vs. $Na_2S_2O_5$) [7].

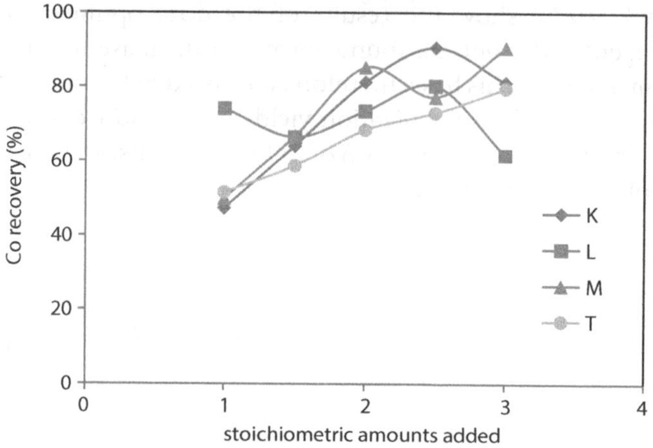

Figure 7.9 Influence of Fe on Co(III) leaching (Co vs. Fe^0) [7].

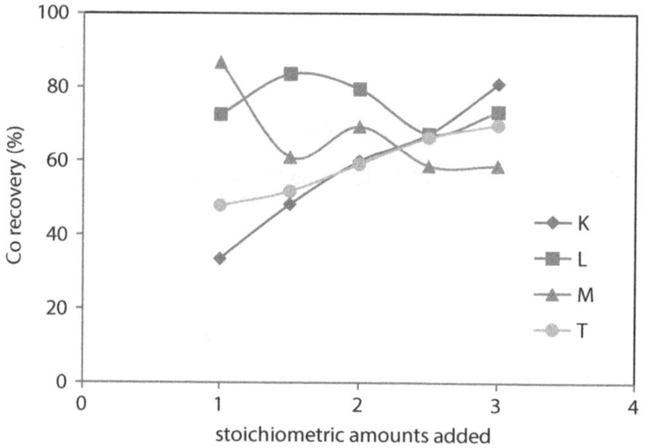

Figure 7.10 Influence of Cu on Co (III) leaching (Co vs. Cu^0) [7].

Overall, it can be seen that after examining all the results relating to the reduction, and therefore, the solubilization of Co^{3+} in each ore with the corresponding reducing agents, that the Kimpe, Luiswishi, and Mukondo ores gave a relatively high yield of 80% on average. On the other hand, the Tilwizembe ore led to a less satisfactory yield of around 70%. This gave us a superficial assessment of the results, which were subsequently deepened with the principal component analysis by highlighting certain details in order to identify the correlated variables, the tests showing similarities or the mineral/reducing associations with atypical behavior.

Following the simultaneous study of the leaching of the four crossed ores with four reducing agents (Fe^{2+}, Fe, Cu, and $Na_2S_2O_5$) of trivalent cobalt, the principal component analysis approach made it possible to highlight the leaching behavior of ores from the copper arc deposit by detecting major process variables and identifying ore or reducing groups with similar metallurgical performances. Optimization tests on the quantities of reducing agents were carried out. The optima retained were crossed with other variables, in particular the Cu and Co yields, reducing agent consumption, Cu and Co solubilization and redox potential at the end of the leaching process.

7.2.4.5 *Microwave Assisted Acid Leaching of Cobalt (III)*

The processing of the mixed oxidized cobalt-copper ores for the recovery of cobalt through sulfuric acid dissolution and electrowinning after impurities removal has necessitated the use of reducing agents and intelligent *in situ* reactions as elaborated in the chapters here above. The present section of the chapter deals with the studies of enhanced dissolution of cobalt bearing minerals as the feed is subjected to microwave energy prior to the leaching. Although the case of microwave assisted leaching, i.e., dissolution while the system is being irradiated with microwave energy, will not be part of the scope of the present section, peculiar aspects of microwaves enhanced comminution prior to the dissolution will be covered. Its industrial importance and relevance, especially in telecommunication, have made the classification of cobalt a critical metal by the European Union (Figure 7.11).

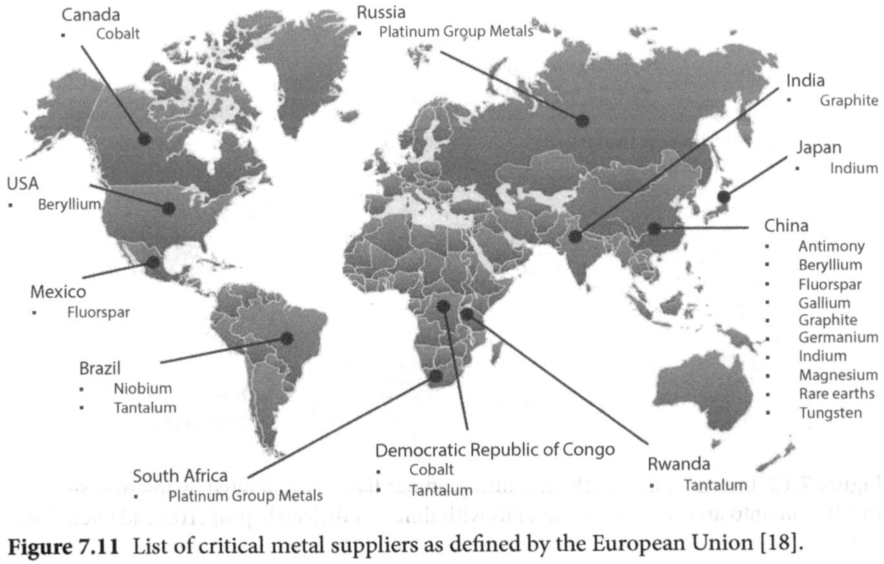

Figure 7.11 List of critical metal suppliers as defined by the European Union [18].

7.2.4.5.1 Background of Microwave-Assisted Comminution

Microwaves as radiations or electromagnetic energy range interact with materials in general, hence with ores according to their dielectric properties. Different minerals in an ore, possessing different dielectric properties will heat differently [19]. The differential or selective heating of the exposed materials leads to differential strains, hence initiating flaws and possibly intergranular cracks and in some situations intragranular flaws, as seen in Figure 7.12.

Similar to a study by Gupta and Denis Yan [20], the ore breakage due to external forces in either impact (e.g., in hammer crusher) or compressive (e.g., in jaw crusher) forces have been modeled using the kinetic model, the matrix model, and the energy model, but the mechanisms of breakage generated by inner forces, resulting from an inner microwave selective heating, is still to be elucidated. Indeed, microwave irradiation of an ore leads to a differential thermal expansion of mixed components or minerals comprising the ore. In the case of a mixed cobalt–copper oxidized ore, depending on the relative composition of the absorbents, opaque and reflective components may lead to an enhancement of the inter-granular fracture of the mixed cobalt–copper ore material. Reported findings on the use of microwaves in minerals processing are numerous [19, 21, 22]. While heating and possible phase transformation are often exploited in synthesis, the differential heating, crack generation, and flaws formation have been exploited in comminution of ores and mineral processing. As an energy range, materials may transmit, reflect, or absorb microwaves. If dipolar, when a material absorbs microwave, the internal dipolar fast and with high frequency aligns itself with

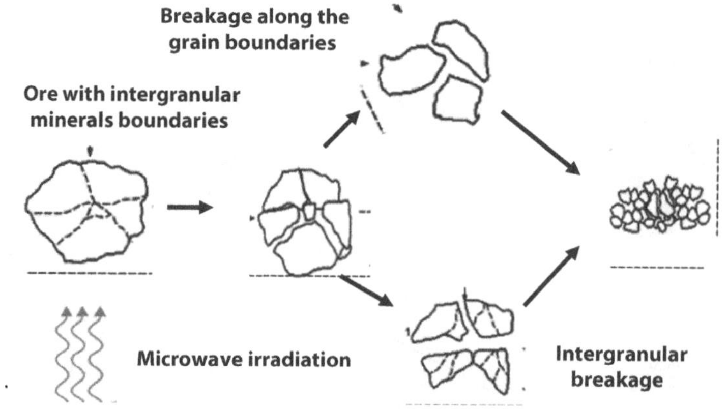

Figure 7.12 Intergranular cracks and intragranular flaws as generated by microwave irradiation onto an ore made of minerals with different dielectric properties, adapted from [20].

the wave's fluctuating magnetic field, hence generating heat through internal friction. The generated heat is dissipated through internal heat transfer leading to (thermal) expansion of the material. The component reflecting and transmitting the waves will be affected at a lesser extent, thus causing a differential heating, which, in turn, will generate internal heat causing stresses to the ore. The intensity of the generated stresses depends on the material's composition, the ratio of its relative components, and its grain size.

7.2.4.5.2 Microwave-Assisted Comminution of High-Grade Cobalt Copper Ore

According to Mulaba-Bafubiandi and Lewis [23], Figure 7.13, the density of the ore is an indicator of the value content. Density tests conducted on a high-grade cobalt copper ore, using the wet displacement method on coarse and fine materials from the studied cobalt–copper mixed oxide ore, led to the fact that the studied materials showed friability and lower densities. Evidence of gas evolution during the wet displacement test was noted, confirming the porousness characteristic of the oxidized ore.

7.2.4.5.3 Microwave Pre-Processing of Cobalt Ore Prior to Leaching

Cobalt bearing minerals, in the complexity of cobalt and copper system, leads to a variety of outcomes depending on the relative Co:Cu ratio. While

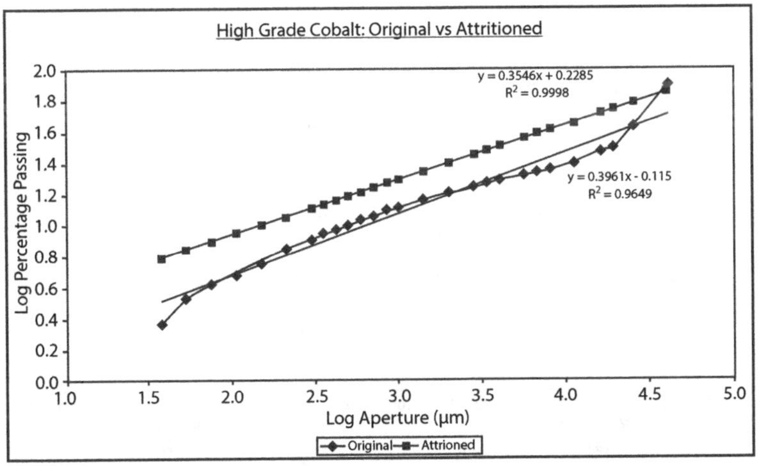

Figure 7.13 Attritioned and original high-grade oxidized cobalt material particle size distribution [23], (with written permission from the SAIMM).

microwave-assisted leaching may be conducted, Figure 7. 14, thus leading to a careful material selection of the leaching container and stirrer, to avoid damaging the magnetron, the section will discuss a pre-processing of the load prior.

Although in the early stage of the usage of microwave energy in mineral processing, multimode cavities were used, the current practice involves the usage of single mode cavities as high and concentrated energy is exploited, effectively speeding the reaction. Figure 7.15(b) shows the metal distribution in fine fractions as generated by a pre-processing of cobalt bearing minerals with a multimode cavity (900 W, 2.45 GHz, 30 min) of cobalt.

Microwaving the cobalt–copper ore leads to up to about 40% Co attritioned, while the non-microwaved showed a value around 30%. The cobalt

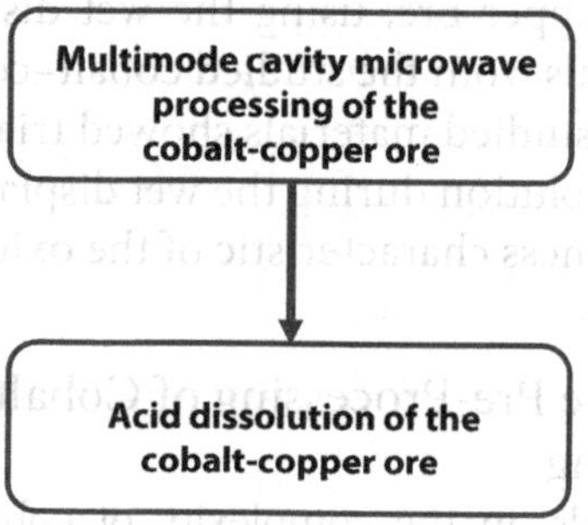

Figure 7.14 The pre-processing of cobalt–copper ore by microwave prior to the dissolution in acidic environment.

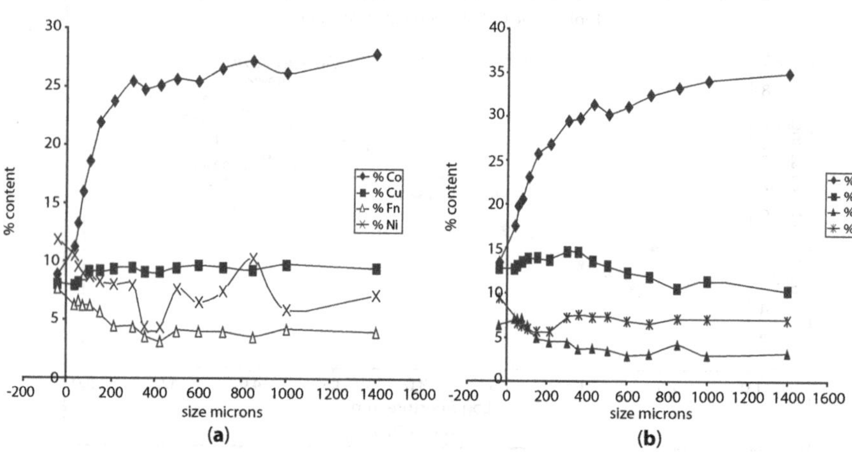

Figure 7.15 (a) Metal distribution in non-microwaved and (b) multimode cavity microwaved cobalt–copper ore [23], (with written permission from the SAIMM).

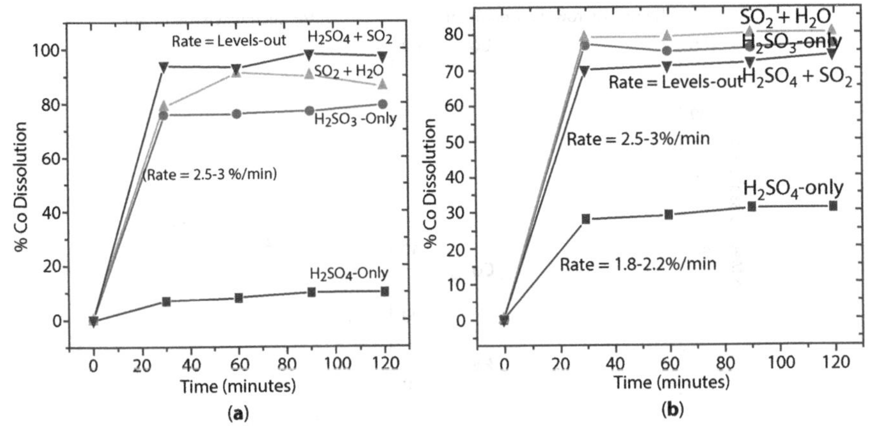

Figure 7.16 Leaching kinetics of cobalt in the high cobalt bearing ore [24], (with written permission from the SAIMM).

variation depicts a logarithmic trend with a correlation coefficient of 91% for the non-microwaved and 99% for the microwaved, respectively.

For other base metals, the dissolution in sulfuric acid of cobalt bearing minerals from a mixed cobalt–copper ore may be enhanced by the use of reducing agents. Often SO_2 is employed. Figures 7.16(a) and 7.16(b) show the kinetics of the dissolution of cobalt bearing minerals in the presence of copper but when SO_2 is employed as a reducing agent. Cobalt dissolution has been found improving irrespective of the Co:Cu ratio when SO_2 is utilized.

The availability of more free acid in the case of a higher ratio would also explain the observed higher dissolution rate.

The generally observed trend is that pre-processing a cobalt–copper oxidized ore with microwave leads to an improved cobalt, as seen in Figure 7.17.

In conclusion, when microwave energy is employed, an improved dissolution of cobalt bearing minerals is observed. The more the cobalt bearing minerals are liberated through microwave pre-processing, the higher the leaching yield will be.

7.2.5 Leaching of Sulfide Minerals

Sulfide ores contain metal sulfides. The sulfur in these ores is at its highest stage-II. As the soluble form of sulfur is SO_4^{2-} at its +VI stage, leaching of sulfide ores will require oxidizing conditions. The reactions of secondary sulfides like CuS and CoS are easier than those of primary sulfides like

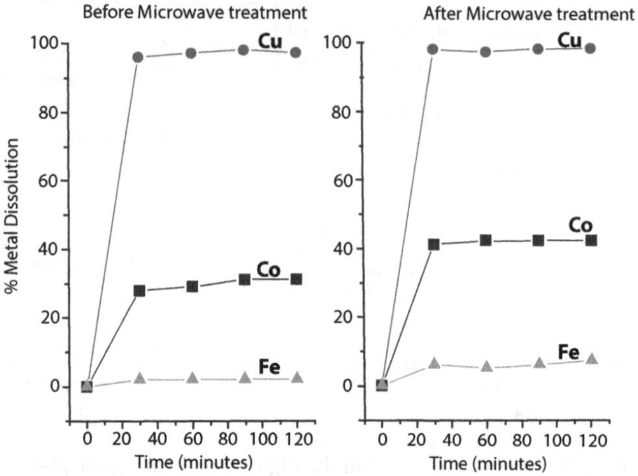

Figure 7.17 Leaching of a mixed complex Co–Cu ore low in Co:Cu ratio [24], (with written permission from the SAIMM).

chalcopyrite $CuFeS_2$. Equations 7.29–7.32 describe these solubilization reactions.

$$CuS + Fe_2(SO_4)_3 \rightarrow CuSO_4 + FeSO_4 + S^\circ \tag{7.29}$$

$$CuS + 2O_2 \rightarrow CuSO_4 \tag{7.30}$$

$$CuFeS_2 + 2Fe_2(SO_4)_3 \rightarrow CuSO_4 + 5FeSO_4 + 2S^\circ \tag{7.31}$$

$$CuCo_2S_4 + 3Fe_2(SO_4)_3 \rightarrow CuSO_4 + 2CoSO_4 + 6FeSO_4 + 4S^\circ \tag{7.32}$$

7.3 Fe^{3+}-Fe^{2+} Redox Cycle in the Dissolution of Mixed Oxidized and Sulfide Minerals

The existence of the Fe^{3+}-Fe^{2+} loop requires a reaction, which oxidizes the Fe^{2+} ions to Fe^{3+} ions and another reaction, which reduces the Fe^{3+} ions to Fe^{2+} ions. The oxidation reactions of Fe^{2+} ions have been widely discussed above in the Co(III) reduction reactions, which require the presence of a reducing agent such as Fe^{2+} ions.

The other reaction is mostly reducing reaction of Fe^{3+} ions by reaction on sulfide. It is described by considering pyrite because it is among the most studied metal sulfide in the literature.

Pyrite oxidation is a complex process involving various oxidation–reduction, complexation, and hydrolysis reactions. It is probably carried out in several steps and involves a certain number of intermediate compounds of variable stability. Pyrite has a reducing character. Sulfur at its minimum stage can only be oxidized. Thus, it passes from the oxidation stage (-II) to (+VI) by taking intermediate values between these two limit values.

However, the authors do not agree on the specific mechanisms involved in these reactions, and many models are proposed. In particular, the nature of the oxidant is controversial. In an aqueous medium, while some claim that dissolved molecular oxygen is the main oxidizing agent, others highlight the role of the ferric ion more precisely [25].

7.3.1 Oxidation by Dissolved Oxygen

Here is one of the oldest models suggested, assuming a prior adsorption of oxygen molecules on the surface of the pyrite:

$$FeS_2 + 2O_2 \rightarrow FeS_2{,}2O_{2(ads)} \tag{7.33}$$

$$FeS_2{,}2O_{2(ads)} \rightarrow Fe^{2+}_{(aq)} + SO^{2-}_{4(aq)} + S^{\circ}_{(s)} \tag{7.34}$$

However, the model does not consider the presence of Fe^{3+} ions encountered in solution. Singer and Stumm [26] proposed a model very close to the previous one, but involving Fe^{3+} ions in three stages:

a) Oxidation of pyrite by dissolved oxygen: During this step, only the sulfur is oxidized into sulfate, the iron being dissolved in the form of Fe^{2+} ions:

$$FeS_2 + \frac{7}{2}O_2 + H_2O \rightarrow Fe^{2+} + SO_4^{2-} + H_2SO_4 \tag{7.35}$$

b) Oxidation of Fe^{2+} to Fe^{3+} by dissolved oxygen:

$$Fe^{2+} + \frac{1}{4}O_2 + H^+ \rightarrow Fe^{3+} + \frac{1}{2}H_2O \quad (7.36)$$

c) Finally, the oxidation of the pyrite by the Fe^{3+} ions previously formed in reaction 7.36.

The oxidation of the sulfur contained in the pyrite exclusively into sulfates (reaction 7.35) has not been clearly established, and some studies mention the possible formation of oxidation products of an intermediate degree: thiosulfates and sulfites. Thus, the mechanism envisaged by Tonazzio [25] is classically initiated by the adsorption of molecular oxygen on the surface of the mineral, but involves various intermediaries such as hydrogen peroxide or thiosulfate ions:

- The reduction of dissolved oxygen adsorbed by water molecule:

$$H_2O + \frac{1}{2}O_2 \rightarrow H_2O_2 \quad (7.37)$$

- Two electrons were transferred to the pyrite by hydrogen peroxide to form water molecule and a metastable intermediate species:

$$FeS_2 + H_2O_2 \rightarrow Fe - S - S - O + H_2O \quad (7.38)$$

- The metastable intermediate specie react with the hydrogen peroxide to form the thiosulfate:

$$Fe - S - S - O + \frac{5}{2}H_2O \rightarrow Fe^{3+} + SO_3^{2-} + OH^- + 2H_2O \quad (7.39)$$

Thiosulfate ions are unstable and quickly oxidize to sulfates. On the other hand, authors consider that the direct oxidation of pyrite by molecular oxygen is improbable because of the magnetic properties of the two compounds [27]. Indeed, the oxygen molecule is paramagnetic while pyrite is diamagnetic. It would, therefore, be the ferric ion that would be the main oxidation agent. This agrees with Breed [28] who observed that no leaching takes place in the absence of Fe^{3+}.

7.3.2 Oxidation by Fe^{3+}

The oxidation of pyrite by ferric ions can be represented by the following equation:

$$FeS_2 + 8H_2O + 14Fe^{3+} \rightarrow 15Fe^{2+} + SO_3^{2-} + 16H^+ \qquad (7.40)$$

This equation is generally considered correct. But it does not provide information on leaching mechanisms.

Research carried out to elucidate these mechanisms clarifies the relative role of dissolved oxygen and Fe^{3+}, and considers that the Fe–S bond is weakened during the process and breaks more easily than the S–S bond. The ferric ion is the main oxidizing agent, but oxidation reactions cannot take place in the absence of dissolved oxygen.

The proposed mechanism is shown in Figure 7.18. The first step is the adsorption of an Fe^{2+} ion on the surface of the pyrite (additional studies having shown that the affinity of Fe^{2+} ions for the surface is much greater than that of Fe^{3+} ions). The Fe^{2+} ion transfers an electron to dissolved oxygen and is oxidized to Fe^{3+}, which then quickly accepts an electron from pyrite. The adsorbed iron is therefore alternately oxidized and reduced. It transfers electrons from pyrite to dissolved oxygen. The oxygen atoms are also transferred from the hydration sphere of the adsorbed iron to the sulfur atoms of the pyrite. The cycle of oxidation or reduction of adsorbed iron and addition of oxygen to the pyrite continues until a stable sulfoxide anion is released from the surface. Depending on the experimental conditions, these anions are sulfates or intermediate oxidized species (thiosulfates or polythionates).

7.3.3 Towards a New "Mineral–Mineral" Process

The oxidation of the pyrite leads to the formation of ferrous ions whether it is done through dissolved oxygen or ferric ions. It is this ferrous ion, which is used in the reduction of Co(III), which in turn is reduced to Co^{2+}, that Fe^{2+} will get oxidized to Fe^{3+}, hence forming a loop for Fe ions.

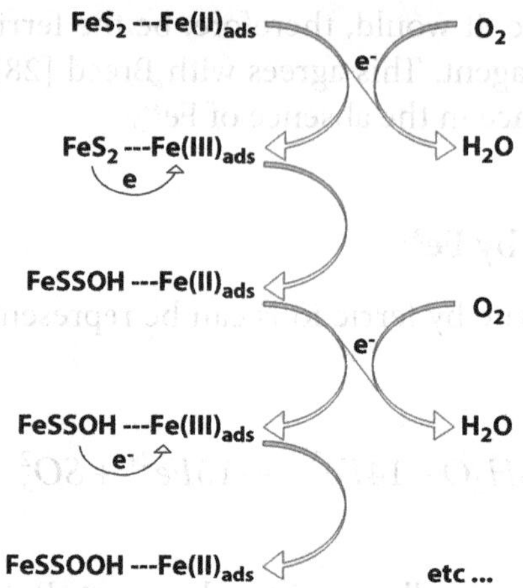

Figure 7.18 Schematic of different reactions involved in pyrite oxidation [25].

Zeka [4] discussed the existence of this loop according to the diagrams in Figure 7.19. The loop for which the existence is proven can be exploited for the benefit of a new Co(III) dissolution process: **mineral–mineral process.**

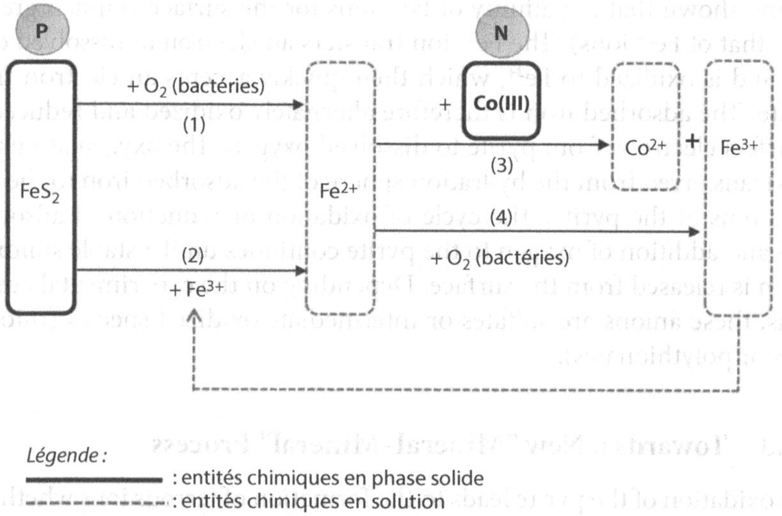

Figure 7.19 Schema showing two oxidation ways of pyrite (P) dissolved oxygen (1) and by Fe^{3+} (2), the two oxidation reactions of Fe^{2+} by cobalt (III) oxide (N(3)) and by dissolved oxygen (4) and the loop of Fe^{3+} recycling [4].

Indeed, as advanced here above, the loop exists when in the reaction medium, there is the simultaneous presence of:

- a reducing agent that consumes the Fe^{3+} ion, in this case, pyrite or any other metal sulfide and
- an oxidant that consumes the Fe^{2+} ion, in this case heterogenite, which contains Co(III).

Zeka [4] also studied the existence of this same loop in the presence of another sulfide and arsenopyrite. In this loop, As(III) and Fe^{2+} simultaneously leached Co(III), and the reaction system can be schematized as shown in Figure 7.20. This system becomes more interesting as arsenopyrite is highly reactive than pyrite.

This iron loop concept introduces a new era in hydrometallurgy and in industrial practices. This development of the Co(III) reduction theory leads to a new process called "mineral–mineral".

This loop was studied by mixing a copper–cobalt ore and a sulfide metal. Other studies have been conducted on the existence of this same loop in a natural system, i.e., on a complex ore comprising both oxidized forms including Co(III) and metal sulfides, which is the subject of the point below.

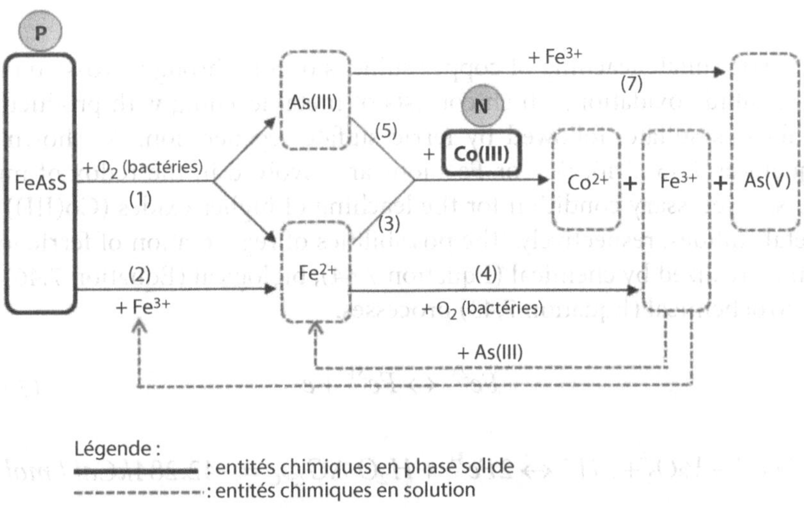

Figure 7.20 Schema of reaction involving the arseno-pyrite (P) [4].

7.4 Application of Mineral–Mineral Leaching Process to the Dissolution of Complex Ores

The existence of the Fe^{2+}-Fe^{3+} loop in the leaching of a mixed system has been studied [3, 5, 16]. Its role in the simultaneous dissolution of Cu and Co from a mineral composite ($CuFeS_2$-Fe_3O_4-Co_2O_3) has been proven [6]. The studied system consists of samples from several origins with specific mineralogies. The first sample is a sulfide ore where the containing chalcopyrite ($CuFeS_2$) is the predominant mineral, in this specific case, the main source of copper and iron. The second sample is an oxide cobalt ore whose heterogenite (CoOOH) is the main source of cobalt. The third sample consists of iron oxides, constituting the iron source and found in the form of magnetite (Fe_3O_4) and limonite ($Fe_2O_3.nH_2O.pFeO(OH)nH_2O$).

The direct leaching of sulfur minerals is essentially an oxidation reaction. The oxidation–dissolution reactions of metal sulfides using the ferric ion are reviewed (Equation 7.41 and Equation 7.42).

$$Cu_2S + 2Fe_2(SO_4)_3 \rightarrow 2CuSO_4 + 4FeSO_4 + S^\circ$$
$$\Delta Gat_{25^\circ C} = -2.147 kCal/mol \tag{7.41}$$

$$CuFeS_2 + 2Fe_2(SO_4)_3 \rightarrow 2CuSO_4 + 5FeSO_4 + S^\circ$$
$$\Delta Gat_{25^\circ C} = -16.108 kCal/mol \tag{7.42}$$

Ferric sulfate leaching of copper sulfides occurs through two-step process: sulfide oxidation, which consists of metal leaching with production of ferrous sulfate, followed by ferric sulfate regeneration. As shown in Equations 7.43–7.46, Fe^{2+} or Fe^{3+} ions are involved in the redox of minerals, a necessary condition for the leaching of higher oxides (Co(III)) or metal sulfides, respectively. The possibilities of regeneration of ferric ions can be realized by chemical (Equation 7.44), biological (Equation 7.46) or electrochemical (Equation 7.43) processes.

$$Fe^{2+} \leftrightarrow Fe^{3+} + \grave{e} \tag{7.43}$$

$$2Fe^{2+} + ½O_2 + 2H^+ \leftrightarrow 2Fe^{3+} + H_2O \; \Delta G_{25^\circ C} = -42.284 kCal/mol \tag{7.44}$$

$$Fe_2(SO_4)_3 + SO_2 + H_2O \rightarrow FeSO_4 + 2H_2SO_4 \ \Delta G_{25°C} = 2.185\ kCal/mol \tag{7.45}$$

$$2Fe^{2+} + \frac{1}{2}O_2 + 2H^+ \xrightarrow{bacteria} 2Fe^{3+} + H_2O \tag{7.46}$$

7.4.1 Reaction Mechanism

The mechanism of the $CuFeS_2$–Fe_3O_4–Co_2O_3 system could be described as follows: the three minerals being present initially in the reaction medium consisting of acidulated water. Initially, the medium was neither oxidizing nor reducing. The process is triggered by the action of sulfuric acid on iron oxides, resulting in the release of Fe^{3+}. The presence of Fe^{3+} makes environment oxidizing, and therefore, is conducive to the dissolution of the copper of the chalcopyrite. This dissolution of copper consequently leads to the reduction of Fe^{3+} to Fe^{2+}, the latter in turn reducing the trivalent cobalt. The Fe^{2+}-Fe^{3+} loop is created.

7.4.2 Redox Test Results of the $CuFeS_2$–Fe_3O_4–Co_2O_3 System

The above redox tests focused on the influence of the mass of iron oxides, Figure 7.21, on the dissolution of copper from chalcopyrite and trivalent

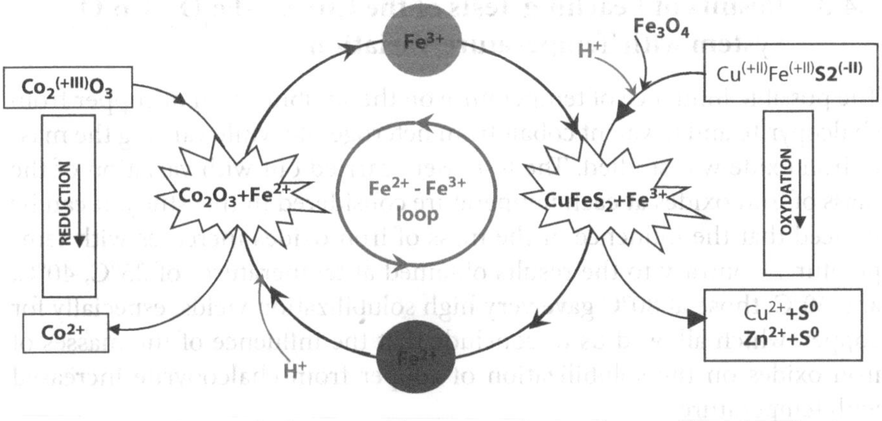

Figure 7.21 Fundamental schematic of the process of mineral–mineral oxydo-reduction [6].

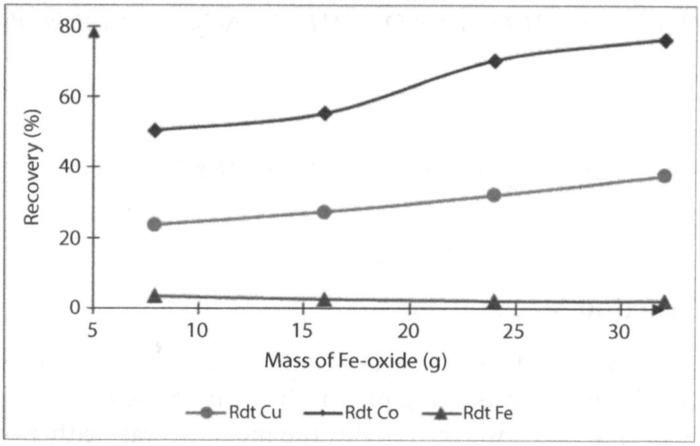

Figure 7.22 Copper, cobalt, and iron recoveries in terms of the mass of iron oxides [6].

cobalt from heterogenite at a temperature of 25°C. Figure 7.22 shows the progression of the solubilization yields of Cu, Co, and Fe at 25°C as a function of the mass.

Varying the iron ore amount shows a slight influence compared to the variation of the acidity, this indicates that the process is chemically controlled. Therefore, for a dependence of chemical reaction, the leaching process would depend on the temperature, to a first approximation, hence the series of simultaneous dissolution tests of Cu and Co with varying temperatures.

7.4.3 Results of Leaching Tests of the $CuFeS_2$–Fe_3O_4–Co_2O_3 System with Temperature Variation

The possible influence of temperature on the solubilization of copper from chalcopyrite and trivalent cobalt from heterogenite while varying the mass of iron oxide was studied. The tests were carried out with variation of the mass of iron oxides at each temperature considered in this study. It can be noticed that the influence of the mass of iron oxides increases with temperature. Contrary to the results obtained at temperatures of 25°C, 40°C, and 60°C, those at 80°C gave very high solubilization yields, especially for copper, which allowed us to conclude that the influence of the masses of iron oxides on the solubilization of copper from chalcopyrite increased with temperature.

7.4.4 Discussions

The leaching tests of the $CuFeS_2$–Fe_3O_4–Co_2O_3 system made possible the study of the conditions of dissolution of copper from chalcopyrite and cobalt from heterogenite.

The iron in solution does not only come from the dissolution of the iron oxides concerned, but also from the leaching of the chalcopyrite in the presence of the ferric ions produced. The influence of the mass of iron oxides on the solubilization of copper from chalcopyrite is perceptible. This influence would be explained by the increased presence of ferric ions as this mass of iron oxide increases, making the environment very oxidizing and favorable to the oxidation of the copper of the chalcopyrite. However, the presence of ferrous ions is essential to maintain the balance of the system because the increased concentration of ferric ions could be detrimental to efficient oxidation (Hiroyoshi, 2008). The presence of the heterogenite kept the two iron sulfates ($Fe_2(SO_4)_3$ and $FeSO_4$) in solution by the formation of the Fe^{2+}-Fe^{3+} loop.

As for the dissolution of the trivalent cobalt of the heterogenite by reducing it to bivalent cobalt, the temperature had very little influence. The strong solubilization could be attributed to the increased presence of ferrous sulfate resulting from the oxidative solubilization of copper from chalcopyrite by the reduction of ferric sulfate at the high iron oxide masses (24 and 32 g) and temperatures (60°C and 80°C). It is observed that the variation in the mass of iron oxides had a great influence on the solubilization and oxidation of the copper of the chalcopyrite and on the reduction of the trivalent cobalt of the heterogenite (Co_2O_3).

The presence of ferric sulfate in solution allows the diffusion of ferric ions through the passive layer, which covers the metal sulfides during their oxidation. Studies have been shown the presence of this layer on chalcopyrite. The diffusion of ferric ions through the passivation layer is accentuated by the rise in temperature.

The influence of temperature on the solubilization of trivalent cobalt from heterogenite by reduction to divalent cobalt is only weak. Cobalt in its Co_2O_3 form is stable. To solubilize it in sulfuric acid, it is necessary to add another suitable reagent that reduces it to bivalent cobalt, which is the soluble form of cobalt in this medium. It is noticed that at iron oxide masses of 24 and 32 g, the solubilization yields of cobalt increase very slightly at temperatures of 60°C and 80°C; temperatures at which large amounts of ferrous sulfate are present, thereby reducing and dissolving more trivalent cobalt.

Acid regeneration was observed. Indeed, the acid consumption shown decreases as the mass of iron oxides increases. At 24 and 32 g of iron oxide at a temperature of 80°C, the acid consumption values by the gangue (GAC) are negative, -10.49 and -6.48 kg/t, respectively, which reflect a regeneration of acid during leaching tests.

Studies have shown that when sulfuric acid has been consumed in the leaching system due to the continued formation of copper and iron sulfates, ferric sulfate begins to undergo hydrolysis according to reaction 7.47:

$$Fe_2(SO_4)_3 + 6H_2O \rightarrow 2Fe(OH)_3 + 3H_2SO_4 \ \Delta G = 49{,}309k\ Cal/mol \tag{7.47}$$

Jarosite precipitation starts at pH = 2 in the presence of suitable monovalent cations. The formation of jarosite, during leaching in the presence of ferric sulfate in the temperature range of 20°C to 100°C, cannot be avoided. The pH at which jarosite is formed decreases when the temperature increases to a pH below 1. Jarosite is formed according to reaction 7.48:

$$\begin{aligned} 3Fe_2(SO_4)_3 + 12H_2O + 2A^+ &\rightarrow 2AFe_3(SO_4)_2(OH)_6 + 5H_2SO_4 \\ &+ 2H^+ \ \Delta G = 41{,}411k\ Cal/mol \end{aligned} \tag{7.48}$$

Where A+ is K^+, Na^+, H_3O^+, Ag^+, $(Pb^{2+})_{0,5}$ …

It is assumed that the presence of at least one of these ions in solution explains acid regeneration. Taking into account two reactions given (7.47) and (7.48), we consider reaction (7.48) to be the most probable because of the reasons mentioned and reaction (7.47) to be the least probable. The system studied made possible certain reactions (Equations 7.47 and 7.48) giving positive ΔG values allowing the simultaneous valorization of copper, cobalt, and iron ores. The latter does not offer any prospect of valuation to date.

7.5 Conclusion

Currently, cobalt is a well sought-after element due to high demand, particularly in the constituent materials of batteries of electrical vehicles. In nature, cobalt is often associated with copper in oxide and sulfide copper-cobalt deposits. Large reserves are reported in the African Copperbelt area stretching between the DRC and Zambia. In the oxidized form of cobalt, its metallurgical treatment relies on a crucial step, namely leaching.

When the cobalt is in the form of trivalent cobalt, its dissolution requires the presence of a reducing agent. In this chapter, all the uses of several reducing reagents have been discussed to present their advantages and disadvantages. The mineral–mineral or solid–solid process, based on bringing a mineral into contact, that plays the role of a reducing agent and consequently extracts an electron from Co(III) to facilitate its leaching was discussed. This eliminates the SO_2 gas emissions generally encountered when metabisulfite is used or SO_2 gas from sulfur roasting is utilized as a reducing agent. One reducing mineral that has been tested is pyrite. The first results show that this is an innovative method. Almost all of the cobalt has gone into solution in this system.

References

1. Mbuya, B., Kime, M., Tshimombo, A., Comparative study of approaches based on the taguchi and anova for optimising the leaching of copper–cobalt flotation tailings. *Chem. Eng. Commun.*, 204, 4, 512–521, 2017.
2. Ntakamutshi, P., Kime, M., Mwema, M., Ngenda, B., Kaniki, T., Agitation and column leaching studies of oxidised copper-cobalt ores under reducing conditions. *Miner. Eng.*, 111, 47–54, 2017.
3. Zeka, L., Lambert, F., Frenay, J., Gaydardzhiev, S., Ilunga, N.A., Possibilities for Co(III) dissolution from oxide ore through simultaneous bioleaching of pyrite. *Miner. Eng.*, 2015.
4. Zeka Mujinga, L., *Lixiviation du cobalt trivalent de l'hétérogénite d'un minerai oxydé cuprocobaltifère du Katanga par le fer ferreux provenant de l'oxydation chimique et bactérienne de la pyrite-arsénopyrite*, p. 346, Thèse de Doctorat, Université de Lubumbashi, 2015.
5. Mwema, M.D., Mpoyo, M., Kafumbila, K., Use of sulphur dioxide as reducing agent in cobalt leaching at Shituru hydrometallurgical plant. *J. South Afr. Inst. Min. Metall.*, 2002.
6. Mbuya, B.I., Ntakamutshi, P.T., Muhungu, S.T., Kawama, F.N., Kasongo, G.K., Nkulu, G.N., Simultaneous leaching of copper and cobalt in the CuFeS2-Fe3O4-Co2O3 system governed by the Fe2+-Fe3+ loop. *Congosciences*, 7, 2, 70–80, 2019.
7. Mbuya, B., Ntakamusthi, P., Kime, M., Zeka, L., Nkulu, G., Mwamba, A., Mulaba-Bafubiandi, A., Metallurgical evaluation of the leaching behavior of copper–cobalt-bearing ores by the principal component analysis approach:

Case study of the drc copperbelt ore deposits. *J. Sustain. Metall.*, 7, 3, 985–994, 2021.

8. Mbuya, B., Kime, M., Ntakamutshi, P., Mwilen, T., Muhungu, S., Mwema, E., Kanda, J., Kaniki, A., Evaluation of flocculation and settling behavior of leach residues: Contribution of principal component analysis. *J. Sustain. Metall.*, 4, 4, 485–492, 2018.
9. Mbuya, B. and Mulaba-Bafubiandi, A.F., Predicting optimized dissolution of selected African copperbelt copper-cobalt-bearing ores by means of neural network prediction and response surface methodology modeling. *Process Integr. Optim. Sustain.*, 2023.
10. Mbuya, B., *Combined Case-Based Reasoning, Artificial Neural Network and Response Surface Methodology for Optimisation of Cobalt and Copper Extraction From Selected Democratic Republic Of Congo Copperbelt Mines*, p. 130, Masters, University of Johannesburg, 2022.
11. Mbuya, B., Meta-Mvita, J., Mulaba-Bafubiandi, A., Bayesian statistics study of a sustainable dissolution of cobalt-bearing minerals from Cu-Co ores. *Can. J. Chem. Eng.*, 1–13, 2023.
12. Kasongo, K. and Mwanat, H., Application of taguchi method and artificial neural network model for the prediction of reductive leaching of cobalt(iii) from oxidised low-grade ores. *S. Afr. J. Sci.*, 117, 5/6, 2021.
13. Mambwe, M., Lavoie, S., Delvaux, D., Batumike, J., Soft sediment deformation structures in the Neoproterozoic Kansuki formation (Katanga Supergroup, Democratic Republic of the Congo): Evidence for deposition in a tectonically active carbonate platform. *J. Afr. Earth Sci.*, 150, 86–95, 2019.
14. Kime, M. and Makgoale, D., Characterization of copper–cobalt ores and quantification of Cu2+, Co2+, Co3+, and Fe3+ in aqueous leachates using uv/visible spectrophotometry. *Chem. Eng. Commun.*, 203, 12, 1648–1655, 2016.
15. Nkulu, G., Gaydardzhiev, S., Mwema, E., Statistical analysis of bioleaching copper, cobalt and nickel from polymetalic concentrate originating from Kamoya deposit in the Democratic Republic of Congo. *Miner. Eng.*, 48, 77–85, 2013.
16. Zeka, L., Frenay, J., Gaydardzhiev, S., Ilunga, N.A., The role of bacteria and the addition of pyrite during leaching of Co(III) from heterogenite. *XXVIII International Mineral Processing Congress–IMPC 2016*, 2016.
17. Ferron, C.J. and Henry, P., The use of ferrous sulphate to enhance the dissolution of cobalt minerals, in: *Hydrometallurgy 2008—6th International Symposium, SME*, Littleton, pp. 1088–1097, 2008.
18. EC DG-ENTR, *Critical Raw Materials For The EU: Report of the Ad-hoc Working Group on Defining Critical Raw Materials*, Version 30 July 2010, Available at: http://ec.europa.eu/enterprise/policies/raw-materials/critical/index_en.htm, 2010.
19. Bradshaw, S.M. and Beckmann, A., Microwave processing of chalcopyrite, in: *Mineral Processing 98*, 1997.

20. Gupta, A. and Yan, D., Chapter 11-Mathematical modelling of comminution processes, in: *Mineral Processing Design and Operations, An Introduction*, Second Edition, pp. 317–355, Elsevier, 2016.
21. Kingman, S.W., Vorster, W., Rowson, N.A., The influence of mineralogy on microwave assisted grinding. *Miner. Eng.*, 13, 3, 313–327, 2000.
22. Patnaik, N. and Rao, R.B., Microwave energy in mineral processing–a review. *IE(I) Mineral J.*, 84, 56–61, February 2004.
23. Mulaba-Bafubiandi, A.F. and Lewis, J.M.L., Effects of microwave energy treatment on value, in: *The South African Institute of Mining and Metallurgy The Third Southern African Conference on Base Metals*, Zimbabwe, Victoria Falls, 2006.
24. Mulaba-Bafubiandi, A.F., Ndalamo, J., Mamba, B.B., Microwave-assisted SO2 flushed acid leaching of mixed cobalt-copper oxidised ores. *Base Metals 2007–'Africa's Base Metals Resurgence'*, SAIMM, 2007.
25. Toniazzo, V., *Approche des mécanismes d'oxydation de la pyrite par Thiobacillus ferroxidans grâce à la modélisation morpho-chimique de la surface minérale: Rôle fondamental des phases superficielles oxydées*, p. 13, 14 et, 17, Thèse. Université Henri Poincaré, 1998.
26. Singer, P.C. and Stumm, W., Kinetics of the oxydation of ferrous iron, in: *Proc. 2nd Symp. Coal Mine Drainage Res*, pp. 12–34, 1968.
27. Moses, C.O., Nordstrom, O.K., Hermann, J.S., Mills, A.L., Aqueous pyrite oxidation by dissolved oxygen and by ferric iron. *Geochim. Cosmochim. Acta*, 51, 1567–1571, 1987.
28. Breed, A.W., Harrison, S.T.L., Hansford, G.S., A preliminary investigation of the ferric leaching of a pyrite/arsenopyrite flotation concentrate. *Miner. Eng.*, 10, 1023–1030, 1997.

8

Rare Earth Elements (REEs) in Complex Ores and Spent Materials: Processing Technologies and Relevance in the Global Energy Transition

Theophilus Ile Ojonimi[1]*, Tina Phiri Chanda[2], Ilemona C. Okeme[3], Ferdinand Asuke[4] and Antoine F. Mulaba-Bafubiandi[5]

[1]Department of Mining Engineering, University of Jos, Jos, Plateau State, Nigeria
[2]Faculty of Energy and Engineering, Murdoch University, Western Australia, Perth, Western Australia, Australia
[3]Interface Analysis Centre, School of Physics, University of Bristol, Bristol City, England, United Kingdom
[4]Department of Metallurgical and Materials Engineering, Ahmadu Bello University, Zaria, Nigeria
[5]Department of Metallurgy, University of Johannesburg, Johannesburg, Gauteng, South Africa

Abstract

Rare earth elements (REEs) as found on the periodic table consist of lanthanides, namely lanthanum, cerium, praseodymium, neodymium, promethium, samarium, europium, gadolinium, terbium, dysprosium, holmium, erbium, thulium, ytterbium, lutetium, and transition metals yttrium and scandium. In view of the rising global campaign for transition to green economy and energy technologies, there has been a corresponding increase in the relevance of these elements since they are required in the manufacturing of key components in clean energy devices such as wind turbines, solar panels, and electric cars. As a result, the gap between the demand and supply of these critical minerals in the global market has widened in recent years. The supply crisis is partly because REEs often exist in complex primary deposits, which would require sustainable economic and environmentally friendly processing to achieve profitability. This book chapter describes rare

**Corresponding author*: t.ojonimi@gmail.com

Elvis Fosso-Kankeu, Bhekie B. Mamba and Antoine F. Mulaba-Bafubiandi (eds.) Recovery of Values from Low-Grade and Complex Minerals: Development of Sustainable Processes, (217–248) © 2024 Scrivener Publishing LLC

earth chemistry, global distribution, their complex ores, possible conventional and environmentally friendly recovery technologies, as well as their relevance in global energy transitions. The authors are fully persuaded that the chapter would be useful to students, teachers, researchers, industry players, and policymakers in the mining and mineral processing industry.

Keywords: Rare earth elements, complex ores, processing technologies, energy transitions

8.1 Introduction

As essential materials requirements for socio-cultural advancement, extracted minerals serve as a conduit for national economic prosperity in the form of increased gross domestic product (GDP) and job creation down the value chain for citizens of nations endowed with such resources. The sector engenders economic diversification as it provides essential raw materials for other sectors of the economy and creates most of the high-paying jobs in a multiplier pattern. An estimated 2.8 additional jobs are generated in other sectors of the economy per metal mining job while 2.5 jobs are additionally generated elsewhere in the economy per non-metal mining job. The rare earth market alone supports about 43,000 jobs in the US, while generating nearly 300 million dollars in revenues [1]. About 82.3 billion dollars' worth of minerals raw materials was produced in the US in 2020, accounting for 14% of the US GDP [2]. Similarly, significant national economic growth recorded in many developing countries is stimulated by the mineral mining sector [3]. These cases should always suffice provided there are minimal issues of resource mismanagement and other forms of widespread corruption. However, in some countries, mining sectors turn out to be conflict flashpoints rather than a blessing in which the overall economic performance does not sufficiently reflect the input of mining [4].

In spite of the associated economic gains, activities in the extractive industry promote extreme environmental pollution, which often outlives the lifespan of mines. This has continued to tarnish the image of the industry, especially in developing countries. Tailings storage facilities constitute one of the greatest threats to the environment aside from the fact that general mining activities directly and indirectly contribute to climate change via emissions of green gases [5]. Therefore, one of the key factors that induce struggles in the path to success of new REEs production projects is the high costs of minimizing the impact of the polluting reagents needed for recovery from their complex ores. In view of this vulnerability, Japan and a few other countries have continued to scout for REEs at seabed of their territorial waters, but no profitable mark has been recorded in the search [6]. Over the last 5,000 years, about 550 million tons of minerals have been extracted. There is a need to extract an equal amount in the next 25 years in order to close the technology's demand and supply gap [*Paraphrased from the words of Lucy Crane, an exploration geologist at Cornish Lithium Limited*]. This remark underscores the need for continuous mining activities in any case. The surest path to a sustainable mining and mineral processing practice is for researchers to continue working at possible cost-effective ways of minimizing the impact of the activities on the environment.

REEs are not scarce; the reason being that Cerium, the most abundant of these Lanthanides occupies the 25th position out of the 78 elements in the earth's crust with an abundance of 60 ppm. Lutetium, which is the least REE, has a crustal abundance of about 0.5 ppm and is 200 times more abundant than gold. However, they are not often concentrated together, which makes their extraction complex and expensive [7]. This has overtime generated a quest for supremacy among nations that are endowed with the ripple effect of limiting supplies to the global market. For instance, in 1990, China became the first country to produce REEs. Owing to the country's huge REE mineral reserves of about 44 million tons, China produces 75% of the global rare earth oxides (RE oxides) from several mines in the country, with the Mongolia region's Bayan Obo accounting for over 50% of light rare earth elements (LREEs) production [7]. Fifteen-thousand metric tons of REEs were mined in the US from its 1.4 million metric tons of reserve in 2018 compared to China's 44 million metric tons. China decreased RE oxides export to Japan where final refining takes place by 34% between the years 2013 and 2017. This probably earned China the current global advantage in the global market. This was achieved partly by cracking down on illegal mining in the country. The US depends on China for its defense industry need for REEs: As of 2018, US 160 million dollars' worth of RE compounds, accounting for 8% of the US need was imported from China [8, 9]. This near monopoly and trade war imposes controlled supply, as well as price instabilities of the materials in the international market [7, 9, 10].

It is on the foundation described above that this chapter has developed subtopics that could be beneficial to students, teachers, researchers, industry professionals, and policymakers in the domains of mining, mineral processing, energy, and environment. The chemistry, occurrence in complex ore bodies, global reserves and distribution, conventional and environmentally friendly processing technologies, and the relevance of REEs in the global energy transitions have been discussed.

8.2 The Chemistry of REEs

REEs are captured as lanthanides on the periodic table with atomic numbers ranging from 57 to 71, namely, lanthanum, cerium, praseodymium, neodymium, promethium, samarium, europium, gadolinium, terbium, dysprosium, holmium, erbium, thulium, ytterbium, lutetium, yttrium, and scandium, all of which are often unevenly distributed on the globe. REEs are normally stable except for Promethium with atomic number 61. Owing to their similarity in chemical and physical properties, Yttrium with atomic number 39 and Scandium with atomic number 21 are grouped with REEs. They are characterized by high luster and typically have indistinguishable silver, silvery-white or gray appearance, which tarnishes with the formation of oxide compounds when exposed to air [11]. The difficulty associated with the separation and concentration of these elements arises from the similarities in their ionic radii, and as such, interchangeable in most minerals. This makes REE ores indeed complex [12, 13]. Generally, a decrease in individual REE concentration is often observed with increase in atomic number, and those with even atomic numbers tend to be more prevalent than REEs with odd atomic numbers in accordance with the Oddo–Harkin's rule, where Ce, La, Nd, Pr, Sm, Gd, Dy, Er, Yb, Eu, Tb, Ho, T m, and Lu occur in ascending order [14–16]. On the basis of their atomic number, REEs are often classified into two groups: the light REEs (LREEs), being La, Ce, Pr, Nd, Sm, and Sc, and the heavy REEs (HREEs), being Eu, Gd, Tb, Dy, Ho, Er, Tm Yb, Lu, and Y [17]). The U.S. Department of Energy named neodymium, terbium, yttrium, dysprosium, and europium as elements critically (CREEs) needed within the next 5 to 15 years, which further classifies the elements in terms of criticality: critical (Nd, Y, Eu, Dy, Tb, and Er), non-critical (Pr, Sm, La, and Gd), and excessive (Tm, Ce, Ho, Lu, and Yb). It is estimated that the average concentration of these elements in the Earth's crust ranges from 130 to 240 μg/g [18, 19].

8.3 REE Minerals and Deposit Types

Rare earth elements are moderately abundant in the Earth's crust and occur in many minerals. However, REEs (excluding the radioactive promethium, which never occurs naturally) are scarcely found in easily exploitable high-grade deposits when compared with their total average crustal abundance of 160 $mg.kg^{-1}$ to 205 $mg.kg^{-1}$. Their mass fraction within minerals and ores is relatively low [20, 21]). Currently, more than 75% of REE production and nearly half of the known reserves (Table 8.1), alongside the majority of REE metallurgical refining and separations occur (or are located) in China. REEs do not occur naturally in the elemental state, but occur as mixtures in several minerals. There are over 200 known minerals containing significant quantities of REEs (most abundant detailed in Table 8.2), with these minerals typically occurring as halides, carbonates, oxides, phosphates, and silicates. Even though there are many RE minerals, a vast majority of

Table 8.1 Rare earth elements' global production and estimated reserves (in tons) [24].

	Mine production		
Country	**2020**	**2021**	**Reserves**
United States	39,000	43,000	1,800,000
Australia	21,000	22,000	4,000,000
Brazil	600	500	21,000,000
Burma (Myanmar)	31,000	26,000	Not applicable
Burundi	300	100	Not applicable
Canada	—	—	830,000
China	140,000	168,000	44,000,000
Greenland	—	—	1,500,000
India	2,900	2,900	6,900,000
Madagascar	2,800	3,200	Not applicable
Russia	2,700	2,700	21,000,000
South Africa	—	—	790,000
Tanzania	—	—	890,000
Thailand	3,600	8,000	Not applicable
Vietnam	700	400	22,000,000
Other countries	100	300	280,000
World total (rounded)	240,000	280,000	120,000,000

Table 8.2 Selected rare earth minerals with their rare earth oxide (REO) % [24, 25].

Mineral	Formula	Approximate rare earth oxide (wt%)
Aeschynite-(Ce)	(Ce,Ca,Fe,Th) $(Ti,Nb,)_2(O,OH)_6$	32
Allanite-(Ce)	$(Ce,Ca,Y)_2(Al,Fe^{3+})_3(SiO_4)_3OH$	38
Apatite	$Ca_5(PO_4)_3(F,Cl,OH)$	19
Bastnaesite-(Ce)	$(Ce,La,)(CO_3)F$	75
Branerrite	$(U,Ca,Y,Ce)(Ti,Fe)_2O_6$	9
Euxenite-(Y)	$(Y,Ca,Ce,U,Th)(Nb,Ta,Ti)_2O_6$	24
Fergusonite-(Ce)	$(Ce,La,Nd)NbO_4$	53
Gadolinite-(Ce)	$(Ce,La,Nd,Y)_2Fe^{2+}Be_2SiO_2O_{10}$	60
Loparite	$(Ce,La,Na,Ca,Sr)(Ti,Nb)O_3$	30
Monazite-(Ce)	$(Ce,La,Nd,Th)PO_4$	65
Parisite-(Ce)	$Ca(Ce,La)_2(CO_3)_3F_2$	61
Synchysite-(Ce)	$Ca(Ce,La)(CO_3)_2F$	51
Xenotime	YPO_4	61
Zircon	$(Zr,REE)SiO_4$	<5

resources (known commercial production and reserves, Table 8.2) are associated with bastnaesite, monazite, and xenotime deposits [21]). Bastnaesite is a Ce-rich REE fluorocarbonate mineral, which is exploited primarily for light REEs (LREEs). It is a readily weatherable mineral found associated with pegmatite and metamorphic zones with REO content of about 70%–75% [22]. Bastnaesite is the primary REE mineral being mined at Bayan Obo, China, and Mountain Pass, USA. Xenotime is a Y-rich REE phosphate mineral, which contains primarily heavy REEs (HREEs) with REO content of about 61% [21]. Xenotime is a chemically stable heavy mineral, a product of both physical and chemical weathering, which get transported and concentrated in placer deposits as detrital minerals. Bastnaesite and xenotime hardly contain minerals naturally occurring radionuclides (NORMs) such as thorium (Th) and uranium (U) [22]. On the other hand, monazite, a phosphate REE mineral rich in LREEs (with REO content of about 65%) is a common accessory mineral of NORMs such as U, Ra, and Th (with varying concentrations and is, therefore, invariably radioactive. Similar to xenotime, monazite is a chemically stable heavy mineral, which is mainly found in placer deposits; notable monazite placer deposits were also found in the study by Gupta and Krishnamurthy [22]. Monazite contains about 4%–12% Th and 0.1–0.3% U. These NORMs constitute a commensurate human health and environmental hazard, extra costs in handling and disposal of wastes generated during REE mining and processing [21, 23]. Other REE minerals known to contain NORMs include fergusonite, euxenite, and aeschynite (Table 8.2).

8.4 REE Ore Mining and Processing Technologies

Mining, comminution, subsequent beneficiation processes of leaching, and refining are the main steps involved in the recovery of rare earth Elements (REEs) from their ores [26]. Depending on the type of mine and ore, rare earth oxides (REO) with concentrations ranging from 0.05% to 10% are often exploited [27, 28]. The methods of mining rare earth (RE) ores may be placer, underground, open pit or *in situ* [29]. Most of the light rare earth elements (LREEs) and heavy rare earth elements (HREEs) are exploited via open pit and *in situ* mining, respectively [30]. Physical beneficiation processes separate the RE minerals from gangue minerals to produce RE mineral concentrates [31]. REO concentrates of approximately 50% is often produced [21]. The beneficiation process for REEs include grinding, gravity separation, magnetic separation, and froth flotation.

The concentrates undergo chemical processing for impurities removal and enrichment of REO content to about 90% [26]. Inorganic acids, such as sulfuric acid (H_2SO_4), hydrochloric acid (HCl), and nitric acid (HNO_3); alkalis (NaOH and Na_2CO_3); and inorganic, acid, and basic salts such as $(NH_4)_2SO_4$, NH_4Cl, and NaCl, are commonly used as reagents in the chemical treatment process [26]. H_2SO_4 and the alkalis are primarily utilized for treatment of phosphate and carbonate–fluoride ores such monazite and bastnaesite [31]. High-temperature acid roasting, which releases fluoride (HF), sulfur dioxide (SO_2), sulfur trioxide (SO_3), and silicon tetrafluoride (SiF_4), is a common practice in China [31]. Capturing of exhaust gases from the process is achieved using a water scrubber, which collects pollutants that were removed using a solvent. While the first scrub recovers HF, H_2SO_4, and H_2SiF_6, the second scrub uses diluted Na_2CO_3 solution to clean the exhaust as it is being emitted [31]. Prior to separation by ion exchange or solvent extraction, the RE minerals, which are fluorocarbonates and phosphates, must be converted to carbonates or chlorides [32].

Various REOs are obtained from the separation process, which often involves solvent extraction [27]. The process utilizes the differences in basicity and the REO content yield can be greater than 99% [21]. Individual REO separation is time-consuming and expensive due to the minor differences in basicity among REEs. Multiple stage solvent extraction is therefore widely used for the separation of REO in trivalent stage, fractionation of REOs into groups and separation of desired individual [21]. Industries commonly use lixiviants such as di-2-ethylhexyl-phosphoric acid (D_2EHPA or P_2O_4) and 2-ethylhexyl phosphonic acid mono-2-ethylhexyl ester (HEH/EHP or P_5O_7) [33, 34]. Ion exchange is a different method for REO separation that can be utilized to produce REOs with purities up to 99.99%, but it is more expensive [32]. High purity REEs are produced via several stepwise processes rather than a single method of purification. Zone melting, molten salt oxide reduction, solid state electrolysis, metallothermic reduction and fused salt electrolysis are some of the final refining processes from which 99.99% REE purity can be achieved [21]. Several REE recovery methods were reviewed and are presented in order to provide a better understanding of modern technologies and their potentials for REE extraction from primary deposits and waste stream, as well as the possibilities of developing new technologies.

8.4.1 Mineral Beneficiation for Recovery of REEs

Beneficiation is for the purpose of enrichment of RE minerals from the ore by means of physical separation. Beneficiation takes advantage of the physicochemical properties of the minerals and these include magnetic separation, which is based on differences in magnetizability; gravity concentration (based on differences in the minerals specific gravities); and froth flotation (based on surface properties of minerals). REEs in bastnaesite (carbonate fluoride), monazite (phosphate), and xenotime (yttrium-phosphate) are enriched into REEs concentrates via beneficiation, which are then treated using hydrometallurgical or pyrometallurgical processes for subsequent separation and concentration of the REEs [35]. Froth flotation is frequently employed for physical beneficiation owing to its capacity to process varying ore particle sizes [36]. Additionally, flotation is the ideal method of beneficiation for complex and fine-grained ores and finds widespread application in Mount Weld ores from Australia [37]. Mineral surface properties can be exploited for froth flotation separation.

Collectors and depressants are common groups of chemicals used in froth flotation. Collectors are reagents, which render the mineral surface hydrophobic, while depressants promote hydrophilicity of gangue mineral particles, thereby inhibiting their flow in the flotation cell. Hydroxamates, fatty acids, dicarboxylic acids, and organic phosphoric acids are typical froth flotation collectors [36]. Sodium silicate, sodium hexafluorosilicate, lignin sulfonate, and sodium carbonate are all common depressants [36]. Removal of the fines is carried out by a desliming stage before flotation. Concentration of monazite can be achieved via density separation owing to its high specific gravity, while magnetic separation methods can be used to extract xenotime from other minerals [38]. Some processes, like gravity separation, have a minimal impact on the environment because they do not require much heating and do not use chemicals.

8.4.1.1 *Recovery of REEs Using Gravity, Magnetic, and Electrostatic Separation*

Small gravimetric differences, slight variation in magnetizability, and surface ionization potential have been exploited for the separation of REEs. Monazite has the highest specific gravity of all the minerals found in beach sand. In decreasing sequence of magnetizability, ilmenite, garnet, xenotime, and monazite exhibit magnetic properties. Ilmenite and rutile exhibit alternating conductivity and nonconductivity during electrostatic separation. Since xenotime has a stronger magnetic field than monazite, it concentrates with ilmenite during magnetic separation. Xenotime, with low electrical conductivity, is separated electrostatically from ilmenite. In the presence of leucoxene, separation of monazite from ilmenite might be complex. Leucoxene's free hematite is transformed into magnetite by a reduction roast at 600°C, making separation simple.

8.4.1.2 *Recovery of REEs from Monazite Using Flotation*

Flotation techniques have been explored for recovery of monazite concentrate from deposits of fine heavy mineral beach sand [39]. Standard gravity methods are not particularly effective when the valuable constituents of heavy mineral deposits are in a finer size range of between 15 and 100 microns, and large losses of the precious finer size constituents have been recorded. The flotation approach is therefore favored [21]. Utilizing fatty acids or phosphoric acid esters as collectors can effectively achieve bulk mineral flotation. The choice of collector depends on how easily the particularly heavy minerals may be removed before separation when using flotation to separate them [21, 39].

8.4.2 Hydrometallurgical Approach for Processing REEs

Hydrometallurgy is an arm of extractive metallurgy that focuses on metals solubilization from their ores, concentrates, or waste streams at relatively low temperatures compared with pyrometallurgy [40]. Hydrometallurgy is a useful technology owing to the fact that it uses appropriate solvents to selectively leach metals from low-grade materials like RE Ores. It is relatively cheap and with low environmental impact [41]. Modern hydrometallurgy is a build-up on Bayer and cyanidation processes of the 18th century. The main primary leaching methods for REE minerals extraction include acid and alkaline. For the acid method, sulfuric acid (H_2SO_4), nitric acid (HNO_3), hydrochloric acid (HCl), and hydrofluoric acid (HF) are exploited for different types of REE concentrates. Sodium hydroxide and sodium carbonate are frequently employed in the alkaline technique.

Rare earths bearing silicates, sulfides, and oxide minerals are the most commonly processed using hydrometallurgical techniques. The aim of hydrometallurgical processes is to separate the various naturally co-occurring REEs from one another, as well as removal of impurity compounds from the material in some circumstances using special operations. Rare earths' chemical characteristics promote facile precipitation and quick dissolution in hydrometallurgical methods. Possibilities for the chemical separation of REEs have been developed by taking advantage of differences in their basicity. The exhibition of stable multivalence in both one of the least plentiful and the most abundant rare earths eases the separation processes.

8.4.2.1 Recovery of REEs from Phosphogypsum

Phosphogypsum (PG) is the by-product of the processing of phosphate ores (fluorapatite, $Ca_5(PO_4)3F$) in acidic medium [42]. The leaching of PG using sulfuric acid (H_2SO_4) for recovery of REEs is the most common method and about 50% of REEs in PG is recovered by leaching [43]. Leaching of PG is followed by solvent extraction or oxalic acid precipitation to recover the REEs from the leachate. A comprehensive method for recovering rare earths from phosphogypsum has been proposed [44]. The procedure involves the use of aqueous H_2SO_4 as lixiviant for leaching of rare earths from phosphogypsum. This is followed by solvent extraction of the elements of interest with the use of nonyl-phenyl phosphoric acid or by selective precipitation with hydrofluoric acid upon the concentration of the leachate via evaporation. Pilot scale study of the technology indicated that it is too complex and expensive for industrial scale application [45]. In order to improve REE dissolution, Hammas-Nasri *et al.* [46] studied REE recovery from PG, utilizing a two-stage H_2SO_4 leaching phase in which H_2SO_4 solutions were used to leach 500 g of PG at a 1:1.3 solid–liquid ratio. REE enrichment of about 86% in the crystallized solid obtained from the evaporation of the second leachate was achieved, which was a mixture of anhydrite–monetite ($CaSO_4$.$CaHPO_4$) phases. In spite of the enrichment of REEs associated with this method, the evaporation process is energy-consuming if it is to be scaled up [47]. Evaporation of the leachate has a positive impact on the separation of REEs from PG, but the technique is unprofitable due to the high production costs associated with heating of the solution. To improve the REE extraction effectiveness using an H_2SO_4 solution, Lokshin *et al.* [48] increased the leaching time to several weeks. In the study, 40 g of PG was leached with H_2SO_4 solutions for 20–30 h using a S:L ratio of 1:2. However, compared to the situation where 1-h leaching was utilized under similar conditions, the leaching efficiency of REEs only increased by three-folds. The recovery of REEs from PG leachates is often achieved using methods that were developed for recovering REE from primary sources. Since the leachate is produced as a mixture of REEs and other metals, techniques like precipitation, solvent extraction, adsorption, and ion exchange have been widely employed for separation and purification [47]. Sulfuric acid has been investigated as a high selectivity leachant for phosphogypsum that dissolves the desired metals, and the method may be simply integrated into an existing phosphoric production plant [41].

The use of HNO_3 as lixiviant in leaching of REEs from PG appears to be the most effective because calcium sulfate is more soluble in the HNO_3 solution when compared with the use of H_2SO_4 and HCl [47]. PG was leached using a 3.0-M HNO_3 and 0.8-M calcium nitrate solutions for 8 h while varying the solid:liquid (S:L) ratios. It was reported that under the same conditions, the leaching efficiency of REEs increased with increasing S:L ratios, but declined as the S:L ratios reduced [49]. A 95% leaching efficiency of REEs was recorded with the use of 15 wt.% HNO_3 concentration and a S:L ratio of 1:10. With a high S:L ratio, it was discovered that the leaching efficiency of REEs rose with HNO_3 concentrations below 15% weight and gradually reduced with concentration increases over 15% weight [50]. However, only a marginal improvement of 6% on the leaching efficiency of REEs and 29% for calcium were achieved by raising the acid concentration from 1.5 to 3.0 M when leaching REEs and calcium ions from PG using S:L ratio of 1:8 acquired at the Agrium fertilizer plant [51].

Despite reports that HNO_3 has a reasonably good leaching efficiency for PG, these techniques are considered unprofitable, especially for industrial scale use [52]. Additionally, because of the limited solubility of PG and the low concentration of REEs to be extracted, total dissolution of PG, in an effort to recover only REEs, is neither possible nor economically feasible [47].

8.4.2.2 Recovery of REEs from Apatite Mineral

Production of phosphoric acid primarily depends on apatite mineral. REEs are often found in the fluorine-laden form of mineral apatite [53]. Equation (8.1) typifies the production of phosphoric acid from $Ca_5(PO_4)_3F$ with the use of H_2SO_4 [54]

$$Ca_5(PO_4)_3F + 5H_2SO_4 + xH_2O \rightarrow 3H_3PO_4 + 5CaSO_4 . x/5H_2O\downarrow + HF\uparrow \tag{8.1}$$

where x is 0 for calcium sulfate anhydrate, 0.5 for hemihydrate, and 2 for dihydrate, i.e., gypsum. The type of calcium sulfate byproduct depends on how much water is used in the process [55]. In the standard procedure, there is always enough water present to create the dihydrate ($CaSO_4.2H_2O$). The apatite is entirely dissolved during the reaction, which transfers all REE to the solution. However, this reaction precipitates insoluble $CaSO_4$ (gypsum), which removes 80% of the REE from the solution [55]. This results from the REE being incorporated during precipitation into the $CaSO_4$ crystal lattice. These REEs are regarded as lost due to the amount of gypsum that is produced under standard processing conditions; 5 tons of gypsum are produced per ton of P_2O_5, according to Zielinski *et al.* [56], as well as the chemical stability of gypsum. This has promoted various approaches for recovery of REE from apatite. The hemihydrate process, in which $CaSO_4.12\ H_2O$ (hemihydrate) rather than gypsum, is produced and precipitated during the digestion of apatite. Unlike gypsum, which only captures 80% of the REEs in the solution, the hemihydrate precipitates virtually all of the REEs in the solution. Hemihydrate is easily leached to recover the REEs [56]. The hemihydrate is filtered upon precipitation and then it is leached with diluted H_2SO_4. This brings the REEs into solution while also dissolving the hemihydrate. Under these circumstances, the REEs in the solution prevent gypsum from re-precipitating, allowing for its removal using solvent extraction [56]. Nitric acid has been shown to be preferable to H_2SO_4 for leaching apatite because it allows for the removal of REEs from the solution prior to its co-precipitating and the easy control of $Ca(NO_3)_2$ solubility [55].

8.4.2.3 *Recovery of REEs from Red Mud*

H_2SO_4 has been found to be the most effective lixiviant for leaching scandium from bauxite. This has been demonstrated with Australian bauxite residue [34]. By combining bis(2-ethylhexyl)phosphoric acid (D2EPHA) and tributyl phosphate (TBP) at a weakly acidic pH of 0.4, scandium could be recovered from the leachate with virtually no co-extraction of iron. An Indian bauxite residue sample containing 110 ppm and 70 ppm of cerium and lanthanum, respectively, was leached using H_2SO_4. The findings demonstrate that a 3-M H_2SO_4 solution produced the highest level of recovery. Maximum recovery of lanthanum was achieved at 35°C, while maximum percentage of cerium occurred at 75°C [57]. During the leaching of bauxite residue with acids, a portion of the acid is used to neutralize the extremely alkaline bauxite residue. After leaching, the rare earths can be extracted from the leachate via subsequent ion exchange, solvent extraction, or selective precipitation technologies [47]. The mineralogical composition of the bauxite residue often dictates the most suitable reagent for leaching and the literature data demonstrate that the acid leaching results greatly depend on the kind of bauxite residue. As a result, a rare earths recovery regime from bauxite residue must be customized for a particular type of bauxite residue. The correlation between the dissolution of iron and the recovery of scandium shows how close the association of scandium and iron minerals is [58]. There is possibility in recovering approximately 50% scandium from bauxite residue samples without introducing an excessive amount of iron into the solution. On the other hand, attempting to recover more than 50% of the scandium will unavoidably cause a significant portion of the iron to dissolve. Without dissolving all of the iron, it is impossible to extract all of the scandium present in the bauxite residue [47].

Since rare earths are associated with iron and titanium minerals that do not really change during the Bayer process used to treat bauxite, they all end up in the red mud (bauxite residue) [47]. Scandium is prevalent in bauxite residue. For instance, Russian bauxite residue was reported to contain 135 ppm of scandium [59], whereas bauxite residue made from Jamaican bauxite had up to 390 ppm of scandium in it [60]. The highest scandium concentrations of up to 1,700 ppm were found in the residue made from bauxites from Moengo in Suriname [61]. These values suggest a large enrichment of scandium in bauxite and bauxite residue as they are significantly higher than the average abundance of 22 ppm of scandium in the Earth's crust. Due to the growing demand for scandium for the preparation of scandium-stabilized zirconia used in solid oxide fuel cells, there is an anticipated increase in demand for the element in the near future [62]. These factors may make the extraction of scandium from red mud of great economic importance [63]. Scandium represents more than 95% of the economic value of REEs in red mud. REEs can be easily extracted from bauxite residue using diluted mineral acids, while most elements like iron cannot be extracted from bauxite under the same experimental conditions [64]. The leaching of bauxite residue using 0.5-M HNO_3 at ambient temperatures and pressures was studied [65]. The findings reveal a high extraction efficiency of 80% for scandium and 96% for yttrium. For the light lanthanides, which provided recovery of up to 50%, the leaching process was less effective. The leaching process was optimized and carried out on a pilot scale in order to recover scandium from bauxite residue [66]. The difficulty of recovering nitrate ions adsorbed to bauxite residue is a problem associated with leaching with HNO_3 [63]. Large volumes of water are produced when there is an excessive amount of water used for flushing, and it is difficult to remove nitrate ions from waste since the majority of nitrate salts are water soluble and cannot be precipitated [41].

By digesting bauxite residue with a dilute acid solution generated by saturating water with SO_2, REEs can be selectively dissolved while leaving the majority of undissolved iron [64]. Additionally, sodalite's sodium, aluminum, and silicon are brought into solution [41]. Solvent extraction can be used to recover rare earths. The residual bauxite can also be leached in an autoclave at temperatures between 125°C and 150°C with a concentrated HCl solution [67]. During this process, iron, aluminum, and REEs all dissolve in the solution. Aluminum chloride hexa-hydrate ($AlCl_3.6H_2O$) can crystallize when the leachate is completely saturated with HCl. Solid $AlCl_3.6H_2O$ is isolated from the solution and calcined into Al_2O_3, with the recovery of HCl gas. The remaining solution is concentrated, and $FeCl_3$ is hydrolyzed to Fe_2O_3 at temperature of about 170°C. After the Fe_2O_3 is removed, a rare earths solution is produced, from which the rare earths can be extracted using a solvent. The highly corrosive nature of HCl gas and HCl solutions is a major problem with an HCl-based process. Leaching with H_2SO_4 is thought to be economically the most effective approach for recovering REEs from bauxite residue, according to recent studies [42, 47, 63].

8.4.2.4 *Recovery of REEs from Calcium Sulfate Sludge*

A method for recovering rare earths from calcium sulfate sludge generated during the production of phosphoric acid from phosphate rock in South

Africa was developed [68]). With 3-M HNO_3, higher leaching efficiency and rare earths maximum recovery were achieved. $Ca(NO_3)_2$ and HNO_3 mixtures were shown to be significantly more efficient than HNO_3-only solutions. For instance, after 72 h of treatment with 1-M HNO_3, a sludge containing 6.8% rare-earth oxides recovered 35% of its rare earth content, whereas the same sludge treated with 1-M HNO_3 and 0.5-M $Ca(NO_3)_2$ recovered 76% of its rare earths. The increase in the leaching efficiency was attributed to the rare-earth ion impurities in the calcium sulfate matrix, which were replaced by calcium ions [68]). An alternate explanation is a decrease in the concentration of free fluoride ions, which results in a rise in the solubility of rare-earth ions [41]. The HNO_3 concentration needed for maximal leaching is decreased from 3 M to 1 M by adding $Ca(NO_3)_2$.

Tributyl phosphate (TBP) or dibutylbutyl phosphonate (DBBP) were used in solvent extraction to remove the dissolved RE ions from the leach solution and DBBP showed the highest yields [68]. The leachate's nitrate concentration has to be increased by adding 2.5-M NH_4NO_3 for effective extraction. NH_4NO_3 might be omitted and the more affordable TBP could take the place of the more expensive DBBP by employing a leachant with a greater concentration of $Ca(NO_3)_2$ (3 M) [41]. Water could be used for stripping. Oxalic acid precipitation and calcination produced a mixed rare-earth oxide. Continuous counter-current tests were conducted on a lab scale, as well as a pilot plant scale to further test the method [69]. Due to the high solubility of rare-earth phosphates and rare-earth alkali metal double sulfates in HNO_3, HNO_3 was shown to be a more effective leachant than H_2SO_4 for recovering rare earths from hemihydrates phosphogypsum [70]. Rare earths recovered reached 96% when a fresh portion of HNO_3 was used, with only 7.8% of the calcium content dissolved into solution. The authors concluded that REEs do not isomorphously co-crystallize with $CaSO_4.12H_2O$, but rather exist as different phases. Grinding the hemihydrates phosphogypsum before leaching with HNO_3 resulted in a greater recovery of rare earths as the grinding operation removed the calcium sulfate crust that covers the surface of the grains of the rare-earth phases [41]. As against the use of H_2SO_4, it has been discovered in another research that leaching phosphate rock with HNO_3 solution completely solubilizes all rare earths as nitrates without producing any solid residues [71]. Except for silicate gangue particles, phosphate rock is easily soluble in 60% HNO_3 at 60°C. After the phosphates have been dissolved, the slurry is filtered to get rid of the silica. By adding $NaNO_3$ to the solution, Na_2SiF_6 can be precipitated to defluorinate the solution. By using ammonia to lower the pH to between 0.3 and 1.4, the rare earths can be recovered, resulting in the formation of a phosphate precipitate that contains 80% to 100% of the rare earths [71]. The rare earths can be extracted by solvent extraction and the process should be completed before crystallization of $Ca(NO_3)_2{\cdot}4H_2O$ [41]. The rare-earth elements can be extracted from the organic phase by first stripping them with water, then precipitating them as hydroxides by adding ammonia to the water phase after that. Controlling pH can be used to achieve selectivity; the highest recovery of rare earths can be attained at a pH of 1.0, where a precipitate first appears [72]. The rare earths remaining in the leachate solution can be recovered using TBP, and the co-extracted H_3PO_4 can be stripped with NH_3.

8.4.2.5 Recovery of REEs from NdFeB Magnet

Most of the RE magnets are neodymium–iron–boron- (NdFeB) based alloys (NdFeB magnets) and samarium–cobalt- (SmCo) based alloys (SmCo magnets). Nd and Sm can therefore be recovered from NdFeB magnets and SmCo magnets, respectively. With the current generation of technology, NdFeB permanent magnets, which contain about 30% REEs, have a variety of engineering applications [73]. NdFeB magnets are made up of a Nd2Fe14B matrix phase and a neodymium-rich grain boundary phase, with small admixtures of praseodymium, gadolinium, terbium, and dysprosium, as well as additional elements like cobalt, vanadium, titanium, zirconium, molybdenum, or niobium [41]. Magnets can have dysprosium added to the grain boundary phase in order to strengthen their temperature stability against demagnetization [41]. Praseodymium, niobium, and dysprosium are vital components of NdFeB permanent magnets, which are commonly used in green energy technologies and consumer electronics owing to their high magnetic performance [73]. The second most popular magnet is SmCo, which has great temperature stability, good corrosion resistance, and a high resistance to demagnetization. $SmCo_5$ or Sm_2Co_{17} are the two different SmCo alloys that are utilized to make permanent magnets. These magnets may also contain admixtures of other transition metals, such as Fe, Zr, and Cu, in addition to Co [41].

The leaching of NdFeB is rather simple because it readily dissolves in mineral acids. The choice of acid is largely determined by the subsequent separation procedure: solvent extraction with HCl and selective precipitation with H_2SO_4 [74]. The use of HNO_3 for leaching magnet scrap is discouraged because it results in nitrated wastewater [55]. The H_2SO_4, is one of the mineral acids commonly used in the leaching process of the NdFeB magnets [75]. Mineral acid leaching is an exothermic process, which enables trivalent REEs to dissolve into solution. In sulfuric acid, neodymium metal dissolves, forming a Nd (III) ion with the release of hydrogen gas [74]). Iron dissolves into the solution as Fe^{2+} or Fe^{3+} ions depending on the type of acid that is used. The recycling of used NdFeB magnets by H_2SO_4 and the precipitation of rare earth compounds using sodium hydroxide (NaOH) and ammonium hydroxide (NH_4OH) were both researched by Lyman and Palmer [75]. To dissolve the magnet particles and maintain a low pH to prevent Fe (II) oxidation to Fe, the weight ratio of acid to scrap was maintained at 2:1 or greater (III). Then, using NaOH and NH_4OH, respectively, REEs were precipitated as sodium sulfate double salt ($Nd_2(SO_4)_3.Na_2SO_4.6H_2O$) and ammonium sulfate double salt ($Nd_2(NH_4)_2(SO_4)_4.H_2O$). The Nd recovery of about 95%, was achieved between the pH values of 1.5 and 2, using NaOH. However, above pH 5, a significant amount of Fe was precipitated. Due to the solubility of Nd salt in the solution in the instance of NH_4OH, the recovery of Nd was also about 70%.

To avoid iron precipitation, Tian *et al.* [76] proposed a method for the recovery of REEs from NdFeB using chelating agents such as tartaric acid and hexamethylenetetramine (HMTA). Uysal *et al.* [73] investigated the dissolution of NdFeB magnet powders in sulfuric acid solution using the Taguchi orthogonal array method. The process defined the optimal process parameters of the NdFeB magnet powders and optimized the influence of sulfuric acid concentration, solid-to-liquid ratio and stirring speed on the extraction of iron and REEs. It was found that the sulfuric acid concentration was a key factor for iron and REE dissolution, while solid-to liquid ratio and stirring speed slightly affected the dissolution of REEs and iron. Maximum iron and REEs extraction were achieved under optimal conditions as the sulfuric acid concentration of 1 M, solid-to-liquid ratio of 1/15 and stirring speed of 350 rpm [73]. Even though no additional heat was provided, the exothermic reactions caused the leaching temperature to rise on its own. In the meantime, leaching SmCo scrap is not too difficult as it can entirely dissolve in 3-M HCl, HNO_3, or H_2SO_4 [73]. Maximum leaching efficiency with minimal acid usage was the target when designing leaching experiments taking environmental issues into account. Processes for SmCo leaching have not been further developed because these magnets have fallen out of favor with the rise of the cheaper and stronger NdFeB magnets. Reactions 8.2 and 8.3 describe how waste NdFeB magnets' primary constituents, Nd and Fe, react when they are dissolved in H_2SO_4 [73].

$$2Nd(s) + 3H_2SO_4(aq) \rightarrow 2Nd^{3+}(aq) + 3SO_4^{-2}(aq) + 3H_2(g) \quad (8.2)$$

$$Fe(s) + H_2SO_4(aq) \rightarrow Fe^{2+}/Fe^{3+}(aq) + SO_4^{-2}(aq) + H_2(g) \quad (8.3)$$

It was found that reactions were more efficient at low temperatures. In order to avoid the need for external heat, leaching procedures were performed at room temperature. Even though no external heat was provided, the exothermic reactions caused the leaching temperature to rise spontaneously. Low temperatures are therefore desirable in the recycling of magnets since leaching efficiency is more important than leaching rate [74].

8.4.3 Pyrometallurgical Approach for Processing of REEs

Pyrometallurgy, unlike hydrometallurgy, may be described as the use of high temperature to partially or completely convert a mineral into elemental form or another chemical compound [77, 78]. Using pyrometallurgical techniques, RE minerals undergo physicochemical changes that result in the separation of REEs. Pyrometallurgical technologies involve a range of processes that can be broadly categorized as calcination, roasting, smelting, and refining. Calcination involves the thermal decomposition of a mineral in the absence of air to remove volatile impurities. Carbonates are often calcined to produce an oxide and CO_2 [79]. In contrast, roasting involves heating a mineral in the presence of air to remove impurities (oxidation). Conversion of sulfides into metal oxide or sulfate and sulfur dioxide is often achieved by roasting [80]. Sulfate roasting, partial roasting, and dead roasting (complete conversion of sulfides into oxides and conversion of sulfides into soluble sulfates) are typical roasting techniques. Smelting involves the melting of concentrates at high temperatures to produce a metal or compound in order to remove impurities. Reducing agents and fluxes are often added to facilitate this process and impurities are removed through the generation of slags, mattes, speiss, and gases [79]. Refining involves purifying the metal produced during previous pyrometallurgical processes. In general, pyrometallurgical processes are often applied in the extraction of metals from high-grade ores; however, their application in the extraction of critical metals from low-grade minerals and processing wastes is limited due to the difficulty associated with treatment of small quantities, high-energy consumption, pollution, and additional costs for environmental remediation [80, 81].

8.4.4 Integrated Pyrometallurgical and Hydrometallurgical Approach for Processing of REEs

Most REEs are often associated with minerals that prevent them from being dissolved in aqueous solutions such as water, acid, or alkaline at room temperature and keep them stable in the solid phase, except for RE ion-adsorbed clay minerals [82]. A roasting pretreatment followed by the use of severe processing conditions can promote the extraction of REEs from REE-containing minerals by releasing the REEs in the concentrates [83]. Salt and oxidation roasting are typical intensification processes that enhance the efficiency of the subsequent acid and alkaline leaching stage [82]. H_2SO_4 baking at temperatures in the range of 250°C–300°C is a widely used method for converting REE minerals to REE sulfate phases, after which the sulfate salts are dissolved in water [84]. Owing to their comparatively low boiling points, other inorganic acids such as hydrochloric acid, hydrofluoric acid, and nitric acid are not ideal for acid baking, but are more suitable for minor REE minerals like eudialyte, apatite, and fergusonite [82]. In subsequent phases, the preferred leaching of REE hydroxides and REE double-sulfate precipitates can also employ the use of hydrochloric and nitric acids.

8.4.4.1 Recovery of REEs from Monazite, Xenotime and Bastnaesite

An integrated approach, which incorporates pyrometallurgy and hydrometallurgy, has also been explored for the treatment of bastnaesite. Bastnaesite concentrate is heated between 400°C and 500°C in a 98% H_2SO_4 solution for a period of time, then the sulfate is leached with sodium chloride containing water [55]. This breaks down the fluorocarbonate with the release of CO_2 and HF gases in the process. REEs are converted during the roasting process into sulfates, which can be leached to selectively precipitate with NaCl. The Bayan Obo mine in China currently employs the sulfuric acid roasting procedure, making it the main technique for bastnaesite processing. REE concentrates can be combined with concentrated H_2SO_4 and roasted at temperatures between 250°C and 300°C for 1 to 3 h [82]. The concentrates react with the H_2SO_4 to form REE sulfates, U and Th sulfates, HF, and CO_2 gases from the bastnaesite. The typical reactions involved during acid roasting are shown in reactions 8.4–8.7 [82]:

$$2REPO_4 + 3H_2SO_4 \rightarrow RE_2(SO_4)_3 + 2H_3PO_4 \tag{8.4}$$

$$2REFCO_3 + 3H_2SO_4 \rightarrow RE_2(SO_4)_3 + 2HF\uparrow + 2CO_2\uparrow + 2H_2O \quad (8.5)$$

$$Th_3(PO_4)_4 + 6H_2SO_4 \rightarrow 3Th(SO_4)_2 + 4H_3PO_4 \quad (8.6)$$

$$2U_3O_8 + O_2 + 6H_2SO_4 \rightarrow 6UO_2SO_4 + 6H_2O \quad (8.7)$$

After roasting, the concentrates are cooled and mixed with water at 30°C - 40°C for 2 to 3 h. Total REE concentration in the leach solution is kept under control at around 60 g/L. Na_2SO_4 salts are added into the leach solution to separate the REEs by forming double-salt precipitates ($(\varkappa RE_2SO_4)_3 . yNa_2SO_4 . zH_2O$) as well as precipitation of Th as a double salt ($Th(SO_4)_2 . Na_2SO_4$) [82]. By adding oxalic acid into the solution, the heavy REE ions that were left in the filtrate double-salt precipitation can be recovered as insoluble REE oxalates. In order to produce REE hydroxides, the mixed double-salt precipitates are treated with a NaOH solution and heated to 95°C for 4–6 h. REEs are preferentially dissolved into the HCl leach solution at equilibrium pH of 3 to 4 after washing the hydroxide mixes with hot water, while Th and Fe hydroxide remain in the residue [82]. The mixed REE chloride solution is delivered to Solvent Extraction (SX), separating the light, middle, and heavy REE intermediate products.

8.4.4.2 *Recovery of REEs Using Alkaline Treatment*

For monazite and xenotime containing phosphate-type minerals, the alkaline treatment process involves caustic soda (NaOH) digestion, as well as sodium carbonate roasting followed by water leaching [42]. In a study by Davris *et al.* [85], the imidazolium ionic liquid 1-ethyl-3-methylimidazolium hydrogen sulfate at 190°C was explored to evaluate the high-temperature leaching of bauxite residue; 70% recovery was recorded for most of the REEs. All the existing iron, titanium, and around 35% of the aluminum were solubilized. The three-step alkaline leach treatment involving reactions (8.8–8.10) has been devised [41]. Hydrochloric acid (HCl) solution is used to dissolve REE carbonate, causing the synthesis of REE fluorides and REE chlorides in reaction 8.8. Using NaOH, reaction (8.9) transforms REE fluorides into REE hydroxide. The generated REE hydroxide is subsequently converted into REE fluorides by reaction (8.10) using HCl.

$$REF_3 . RE_2(CO_3)_3(s) + 6HCl \rightarrow REF_3(s) + 2RECl_3(s) + 3H_2O + 3CO_2 \quad (8.8)$$

$$REF_3(s) + 3NaOH \rightarrow RE(OH)_3(s) + 2NaF \quad (8.9)$$

$$RE(OH)_3(s) + 3HCl \rightarrow RECl_3 + 3H_2O \quad (8.10)$$

8.4.5 Alternative Technology for Processing of REEs

Phytomining and solvometallurgy are two innovative metal extraction techniques that are currently being developed.

8.4.5.1 Phytomining for Production of REEs

Phytomining involves the use of specific hyperaccumulator plants to extract high-value metals from unconventional resources, such as low-grade metal soils and mining waste material. Hyperaccumulators, herein, refer to plants with the capacity to accumulate extremely high concentrations of certain metals in their shoots [86–88]. This method involves choosing hyperaccumulator plants with desirable features, such as high biomass, and growing them over unprofitable ore bodies, mining waste, or metal-contaminated soils. The biomass is then harvested and processed after harvest to create a commercially viable, high-grade bio-ore [89]. When compared to other plants growing in the same soils, hyperaccumulator plants can naturally absorb extraordinary (>100-fold greater) amounts of metal (loids) into their shoot biomass [89]. For a variety of metal (loids), including As, Se, Mn, Co, Ni, Zn, REEs, and Tl, hyperaccumulators are well-known. When choosing the best candidate species, biomass production and shoot metal concentrations are crucial considerations. The biomass is often burned after harvesting to create a "bio-ore" that is used in the product recovery process [87, 88]. The product recovery process initially concentrated on employing pyrometallurgy to extract crude metal from the "bio-ore" [86]. Considering the high purity of the "bio-ore," it is more profitable to extract higher value products such as precious metals rather than crude metal, even though it is subject to the fluctuating market price of the metal [80]. The key steps in developing sustainable phytomining are (1) identifying potential metal-rich soils; (2) selecting suitable hyperaccumulator plant species; (3) testing the agronomy system; and (5) evaluation of possible metal recovery methods [86, 87].

There are 22 plant species that are known to be REE hyperaccumulators worldwide, and REE hyperaccumulation is defined as when plants have more than 1,000 g g^{-1} dry weight total REE in their shoots [90]. The phenomenon of REE hyperaccumulation appears to be unpredictable.

According to the theory, plants that hyperaccumulate Al^{3+}, Si^{3+}, and/or Mn^{3+} have the ability to hyperaccumulate REE. REE uptake is thought to follow from these mechanisms. In China, REE phytoextraction from the well-known REE metal crop *Dicranopteris linearis* is being extensively investigated [90]. However, it is still completely unknown what physiological pathways lead to increased REE uptake and accumulation [37].

The first crucial processes in both procedures are grinding and drying the biomass. The earlier recovery method depends on an improved leaching step that is done in a 0.5-M nitric acid solution. REE purification is carried out by percolating three solutions consecutively through a resin bed after the ion-exchange resin has been placed onto a column: two washing steps with water and 0.75-M nitric acid to remove competing ions and one elution step using 3-M nitric acid. These processes result in 78% REE recovery and 80% REE purity [91]. The second recovery method entails three steps: (1) extraction of REEs from the dried and powdered biomass by leaching with a solution comprising an acid (such as sulfuric acid) and/or complexing agents; (2) REE precipitation with oxalate; and (3) calcination to prepare REO [37]. These procedures yield > 80% REE purity and recover a solid containing 52% REE, resulting in a 2500 REE concentration factor from soil to final solid [92]. The mining industry has not yet tested that theory on a large scale, despite the fact that it has received scientific support during the past few decades [37]. Instead of requiring mine site rehabilitation at the end of its useful life, phytoextraction has an environmental impact similar to that of agriculture or agroforestry [93]. With the added benefit of *in-situ* restoration, phytoextraction is anticipated to revolutionize the extraction of those elements that are inaccessible by conventional mining processes [82].

8.4.5.2 *Solvometallurgy*

Although aqueous solutions of mineral acids are commonly used to leach metal values from phosphogypsum, it is feasible to largely suppress the aqueous phase and utilize organic solvents instead. This technique is known as solvometallurgical process. The solvometallurgical method enables the production of metals with high selectivity, while minimizing the volumes of leaching solutions and acid consumption [41, 94]. Although solvometallurgical techniques were developed in the 1950s and early 1960s for the recovery of uranium from low-grade complex ores, sadly, these techniques have received little attention ever since despite having the potential to recover metals from ores and the waste products of industrial processes [41]. Solvent leaching and slurry solvent extraction are two alternatives that can be used. In solvent leaching, a complexing agent that has been dissolved in an organic solvent is used to carry out the leaching process [41]. The extractant and the metal complex must both be soluble in the organic solvent, which can be miscible or immiscible with water. The metals can be extracted from the organic phase after leaching. In slurry solvent extraction, a little amount of acid solution is used to wet the finely crushed ore. In slurry solvent extraction, the finely crushed ore is contacted with a water-immiscible organic phase that contains an extractant after being wetted by a small volume of acid solution [41]. The volume of the aqueous phase is substantially smaller in this method than in conventional solvent extraction.

Exploration of simultaneous recovery of REEs and removal of radioactive elements (radium, uranium, thorium, cadmium, and chromium) from PG using an organic solution of tri-n-butyl phosphate (TBP), tri-n-octylphosphine oxide (TOPO), was conducted [95]. The results showed that when leaching PG with 0.5-M TBP using a solid:liquid ratio of 1, for 2 h at 55°C, leaching efficiency for radioactive metals ranged from 66% to 73%, and 69% for all REEs. The removal efficiencies of both radioactive elements and REE metals were found to rise with two subsequent leaching processes, but not with a third. However, a significant rise was observed in the case of subsequent steps utilizing a mixture of organic solutions, TBP and TOPO with molar concentrations of 0.7 and 0.9, respectively. Further research revealed that the extractability of these elements was significantly affected by pre-treating PG with hot sodium carbonate prior to leaching with a mixture of TBP and TOPO in kerosene, with an optimal removal efficiency of 95% and 80% being achieved for radioactive elements and REE recovery, respectively.

8.5 Relevance of REEs in Energy Transition

Since the inception of the International Energy Agency (IEA) in 1974, which was necessitated by the economic downturn experienced globally as a result of the severe perturbations in the global oil and gas market, the organization has focused on driving campaigns for sustainable energy options [96]. Energy experts and other stakeholders would think that the lessons learned from the event should now place the global energy sector in the need for transition to clean energy. In spite of this, economic growth across several industrialized nations in the past four decades was achieved by a corresponding growth in coal energy consumption, especially in the manufacturing sector. Perhaps for lack of political will, over the past two decades, the use of hydrocarbons has either not significantly changed or has been on the increase. Most nations of the world have entered climate protection accords only on paper [97]. As of 2017, primary energy consumed 64% coal, while oil and natural gas and renewables, respectively, contributed 20% and 8% to the mix. By 2021, 70% of electricity is still being generated from coal and gas, with renewables accounting for 2.5% [98–100]. New coal-fired power plants are being constructed to support local manufacturing in the African Union (AU) and several other countries in Asia, while some of the old plants are running in the United States and parts of the EU. This poses a problem for the efforts being made by a few nations and companies alike to lower their greenhouse gas emissions to net zero, which are all geared towards the realization of the goals of the UN's clean energy transition programs. This is why global energy policy-making bodies like IEA are interested in critical minerals and their role in clean energy transitions. In order to realize the decarbonization goal, there would be a deployment of a broad spectrum of clean energy technologies, many of which largely depend on critical raw materials. When then can raw materials be said to be critical?

On the basis of supply and recycling rate, an EU raw materials research group carried out a study in 2020. The report defined raw materials as critical when their unavailability or limited supply impacts the economy more than the shortage in supply of other materials. Among the critical materials listed are REEs. For this reason, the global demand for REEs is expected to rise above 40,000 tons per annum [101]. REEs are used in smart phones, electric vehicles, and wind power installations. Table 8.3 provides information about REE applications in some clean devices.

Table 8.3 General application of REEs in clean energy technologies [102].

Application	Sc	Y	La	Ce	Pr	Nd	Pm	Sm	Eu	Gd	Tb	Dy	Ho	Er	Tm	Yb	Lu
Batteries			▲	▲	▲	▲	▲				▲						
Electronics		▲	▲	▲	▲	▲											
Lamps	▲	▲	▲	▲	▲	▲		▲	▲	▲	▲	▲	▲	▲	▲	▲	
Lasers	▲	▲	▲	▲	▲	▲	▲	▲	▲	▲	▲	▲	▲	▲	▲	▲	▲
Magnets				▲	▲	▲		▲	▲	▲			▲				

8.6 Conclusion

Mining is always associated with environmental and social issues. Proffering medium- and long-term solutions to the problem will be beneficial for the production of rare earth elements and their contribution to energy transition technologies. Inclusion of remediation or pollution-mitigating strategies at the infant stages of mines can support conformity with environmental standards throughout the project.

The inclusion of host communities in the utilization of mineral sales proceeds has always constituted a rich field of controversy, especially in countries where artisanal and small-scale mining commonly take place. The authorities concerned must be accountable to the people. This is where mining policies and legislations must be brought to bear.

The surest path to a sustainable mining and mineral processing practice is for researchers to continue working at possible cost-effective ways of minimizing the impact of the activities on the environment. Wastes from clean energy devices such as batteries can provide a platform for future recycling of REEs. This will minimize the pressure on demand for the elements from primary deposits.

References

1. National Mining Association, Economic growth, 2021. https://mineralsmakelife.org.economic-growth [Accessed July 14, 2022].
2. United States Geological Survey, *Mineral Commodity Summaries 2021*, https://www.ugs.gov/centers/nmic/mineral-commodity-summaries. [Accessed July 14].
3. Ericsson, M. and Lof, O., Mining's contribution to national economies between 1996 and 2016. *Miner. Econ.*, 32, 223–250, 2019.
4. Mahonye, N. and Mandishara, L., Mechanism between mining sector and economic growth in Zimbabwe, is it a resource curse? *Environ. Econ.*, 6, 3, 81–92, 2015.
5. Azadi, M., Northey, S.A., Ali, S.H., Edraki, M., Transparency on greenhouse gas emissions from mining to enable climate change mitigation. *Nat. Geosci.*, 13, 100–104, 2020.
6. Stratfor, The geopolitics of rare earth elements, 2019. https://worldview.stratfor.com/article/geopolitics-rare-earth-elements. [Accessed July 14, 2022].
7. Van Gosen, B.S., Verplanck, P.L., Seal II, R.R., Long, K.R., Gambogi, J., Rare-earth elements, in: *Critical Mineral Resources of The United States-Economic And Environmental Geology And Prospects For Future Supply*, K.J. Schulz, J.H.

Deyoung Jr., R.R. Seal II, D.C. Bradley, (Eds.), pp. 01–031, U.S. Geological Survey Professional paper 1802, 2017, https://doi.org/10.3133/pp18020.

8. United States Geological Survey, Rare earths statistics and information, 2018. https://www.usgs.gov/centers/national-minerals-information-center/rare-earths-statistics-and-information. [Accessed July 14, 2022].
9. Taggart, R.K., Hower, J.C., Dwyers, G.S., Hsu-Kim, H., Trends in the rare earth content of us-based coal combustion fly ashes. *Environ. Sci. Technol.*, 50, 5919–5926, 2016.
10. Rybak, A. and Rybak, A., Characteristics of some selected methods of rare earth elements recovery from coal fly ashes. *Metals*, 11, 1, 142, 2021.
11. Connelly, N.G., Damhus, T., Hartshorn, R.M., Hutton, A.T., *Nomenclature of Inorganic Chemistry–IUPAC Recommendations 2005*, vol. 27, pp. 25–26, Chemistry International–Newsmagazine for IUPAC, 2005, https://doi.org/10.1515/ci.2005.27.6.25.
12. Hartley, F.R., *Symposium on Uranium in Australia*, The AusIMM, Rum Jungle, Northern Territory, Australia, 1968.
13. Yang, X., *Leaching Characteristics of Rare Earth Elements From Bituminous Coal-Based Sources*, p. 49, Theses and Dissertations-Mining Engineering, 2019, https://uknwoledge.uky.edu/mng_etds/49.
14. Laveuf, C. and Cornu, S., A review on the potentiality of Rare Earth Elements to Trace pedogenetic processes. *Geoderma*, 154, 1-2, 1–2, 2009.
15. Wang, L., Huang, X., Yu, Y., Long, Z., Kinetics of rare earth pre-loading with 2-ethylhexyl phosphoric acid mono 2-ethylhexyl ester [HEH (EHP)] using rare earth carbonates. *Sep. Purif. Technol.*, 122, 490–494, 2014.
16. Wang, L. and Liang, T., Geochemical fractions of rare earth elements in soil around a mine tailing in Baotou, China. *Sci. Rep.*, 5, 12483, 2015.
17. Moldoveanu, G.A. and Papangelakis, V.S., Recovery of rare earth elements adsorbed on clay minerals: II. Leaching with ammonium sulfate. *Hydrometallurgy*, 131–132, 158–166, 2013.
18. Xiao, Y., Liu, X., Feng, Z., Huang, L., Chen, Y., Wang, L., Long, Z., Recovery pf rare earths from weathered crust elution-deposited rare earth ore. *J. Rare Earths*, 33, 5, 545–552, 2015.
19. Yang, X.J., Lin, A., Li, X.L., Wu, Y., Zhou, W., Chen, Z., China's ion adsorption rare earth resources, mining consequences and preservation. *Environ. Dev.*, 8, 131–136, 2013.
20. Wedepohl, K.H., The composition of the continental crust. *Geochim. Cosmochim. Acta*, 59, 1217–1232, 1995.
21. British Geological Survey, Rare earth element. https://www2.bgs.ac.uk/mineralsuk/download/mineralProfiles/rare_earth_elements_profile.pf [accessed 10 May 2022].
22. Gupta, C.K. and Krishnamurthy, N., *Extractive Metallurgy of Rare Earths*, CRC Press, Boca Raton, 2005.
23. Kanazawa, Y. and Kamitani, M., Rare earth minerals and resources in the world. *J. Alloys Compd.*, 408-412, 1339–1343, 2006.

24. United States Geological Survey, Mineral commodity summaries, January 2022. https://pubs.usgs.gov/periodicals/mcs2022/mcs2022-rare-earths.pdf (accessed 5 May 2022).
25. Weng, Z.H., Jowitt, S.M., Mudd, G.M., Haque, N., Assessing rare earth element mineral deposit types and links to environmental impacts. *Trans. Institute Min. Metallurgy, Section B*, 122, 83–96, 2013.
26. Navarro, J. and Zhao, F., Life-cycle assessment of the production of rare-earth elements for energy applications: A review. *Front. Energy Res.*, 2, 45, 1, 2014.
27. Fuerstenau, D.W., Design and development of novel flotation reagents for the beneficiation of Mountain pass rare-earth ore. *Miner. Metall. Process.*, 30, 1–9, 2013.
28. Golev, A., Scott, M., Erskine, P.D., Ali, S.H., Ballantyne, G.R., Rare earths supply chains: Current status, constraints and opportunities. *Resour. Policy*, 41, 52–59, 2014.
29. Walters, A., Lusty, P., Chetwyn, C., Hill, A., *Rare Earth Elements*, British Geological Survey (BGS), Keyworth, 2010.
30. Wang, Z., Dai, S., Zou, J., French, D., Graham, I.T., Rare earth elements and yttrium in coal ash from the luzhou power plant in Sichuan, Southwest China: Concentration, characterization and optimized extraction. *Int. J. Coal Geol.*, 203, 1–14, 2019.
31. Zhang, J. and Edwards, C., Mineral decomposition and leaching processes for treating rare earth ore concentrates. *Can. Metall. Q.*, 52, 243–248, 2013.
32. McGill, I., *Rare Earth Elements*, Ullmann's Encyclopedia of Industrial Chemistry, Weinheim, 2000.
33. Chang, H., Mei, L., II, Zhaogang, L., II, Yanhong, H.U., Fushun, Z., Study on separation of rare earth elements in complex system. *J. Rare Earths*, 28, 116–119, 2010, doi: 10.1016/S1002-0721(10)60270-0.
34. Wang, L.S., Long, Z.Q., Huang, X.W., Yu, Y., Cui, D.L., Zhang, G.C., Recovery of rare earths from wet-process phosphoric acid. *Hydrometallurgy*, 101, 41–47, 2010.
35. Habashi, F., Extractive metallurgy of rare earths. *Can. Metall. Q.*, 52, 224–233, 2013.
36. Jordens., A., Cheng, Y.P., Waters, K.E., A review of the beneficiation of rare earth element bearing minerals. *Miner. Eng.*, 41, 97–114, 2013.
37. Vaughan, J., Tungpalan, K., Parbhakar-Fox, A., Fu, W., Gagen, E.J., Nkrumah, P.N., Southam, G., van der Ent, A., Erskine, P.D., Gow, P., Valenta, R., Toward closing a loophole: Recovering rare earth elements from uranium metallurgical process tailings. *JOM*, 73, 1, 39–53, 2021.
38. Krishnamurthy, N.C. and Gupta, K., *Extractive Metallurgy of Rare Earths*, 2nd ed, CRC Press, Boca Raton, 2016.
39. Ferron, C.J., Bulatovic, S.M., Slater, R.S., Beneficiation of rare earth oxide minerals, in: *Proc. Int. Conf. Rare Earth Minerals and Minerals for Electronic*

Uses, Hat Yai, Thailand, Jan. 1991 Prince of Songkla University, Thailand, pp. 251–269, 1991.
40. Krishnamurthy, N. and Gupta, C.K., *Extractive Metallurgy of Rare Earths*, CRC Press, Boca Raton, FL, 2004.
41. Binnemans, K., Jones, B., Blanpain, T., Van Gerven, Y., Pontikes, Y., Towards zero waste valorisation of rare-earth-containing industrial process residues: A critical review. *J. Clean. Prod.*, 99, 17–38, 2015.
42. Cánovas, C.R., Chapron, S., Arrachart, G., Pellet-Rostaing, S., Leaching of rare earth elements (REEs) and impurities from phosphogypsum: A preliminary insight for further recovery of critical raw materials. *J. Clean. Prod.*, 219, 225–235, 2019.
43. Peelman, S., Sun, Z.H.I., Sietsma, J., Leaching of rare earth elements, in: *Rare Earths Industry*, pp. 319–334, 2016.
44. Jarosinski, A., Kowalczyk, J., Mazanek, C., Development of the Polish Wasteless Technology of Apatite Phosphogypsum Utilization with Recovery of Rare-Earths. *J. Alloys Compd.*, 200, 147–150, 1993.
45. Wang, H., Kuo, W., Lin, C., Chen, P., Study of the material properties of fly ash added to oyster cement mortar. *Constr. Build. Mater.*, 41, 532–537, 2013.
46. Hammas-Nasri, I., Horchani-Naifer, K., Férid, M., Barca, D., Rare earths concentration from phosphogypsum waste by two-step leaching method. *Int. J. Miner. Process.*, 149, 78–8, 2016.
47. Mukaba, J.L., Eze, C.P., Pereao, O., Petrik, L.F., Rare earths' recovery from phosphogypsum: an overview on direct and indirect leaching techniques. *Minerals*, 11, 1051, 2021.
48. Lokshin, E., Tareeva, O., Elizarova, I., A study of the sulfuric acid leaching of rare-earth elements, phosphorus, and alkali metals from phosphodihydrate. *Russ. J. Appl. Chem.*, 83, 958–964, 2010.
49. El-Reefy, S., Nayl, A., Aly, H., Leaching and group separation of lanthanides from phosphogypsum, in: *Proceedings of the 9th. International Conference for Nuclear Sciences and Applications*, Sharm Al Sheikh, Egypt, pp. 11–14, 2008February 11–14, 2008.
50. Liu, H., Liu, H., Nie, C., Zhang, J., Steenari, B.M., Ekberg, C., Comprehensive treatments of tungsten slags in China: A critical review. *J. Environ. Manage.*, 270, 110927, 2020.
51. Walawalkar, M., Nichol, C.K., Gisele Azimi, G., Process investigation of the acid leaching of rare earth elements from phosphogypsum using HCl, HNO_3 and H_2SO_4. *Hydrometallurgy*, 166, 195–204, 2016.
52. Salo, M., Knauf, O., Mäkinen, J., Yang, X., Koukkari, P., Integrated acid leaching and biological sulfate reduction of phosphogypsum for REE recovery. *Miner. Eng.*, 155, 106408, 2020.
53. Preston, J.S., Cole, P.M., Craig, W.M., Feather, A.M., The recovery of rare earth oxides from a phosphoric acid by-product. Part 1: Leaching of rare earth values and recovery of a mixed rare earth oxide by solvent extraction. *Hydrometallurgy*, 41, 1–19, 1996.

54. Wang, W., Pranolo, Y., Cheng, C.Y., Metallurgical processes for scandium recovery from various resources: A review. *Hydrometallurgy*, 108, 100–108, 2011.
55. Peelman, S., Sun, Z., Sietsma, J., Yang, Y., Hydrometallurgical extraction of rare earth elements from low grade mine tailings, in: *Rare Metal Technology 2016*, S. Alam, H. Kim, N.R. Neelameggham, T. Ouchi, H. Oosterhof (Eds.), pp. 17–29, John Wiley & Sons, 2016, https://doi.org/10.1002/9781119274834.ch2.
56. Zielinski, S., Szczepanik, A., Buca, M., Kunecki, M., Recovery of lanthanides from kola apatite in phosphoric acid manufacture. *J. Chem. Technol. Biotechnol.*, 56, 4, 355 360, 1993.
57. Abhilash, Meshram, P., Sarkar, S. *et al.*, Exploring blast furnace slag as a secondary resource for extraction of rare earth elements. *Min. Metall. Explor.*, 34, 178–182, 2017. https://doi.org/10.19150/mmp.7857.
58. Borra, C.R., Pontikes, Y., Binnemans, K., Van Gerven, T., Leaching of rare earths from bauxite residue (red mud). *Miner. Eng.*, 76, 20–27, 2015.
59. Reid, S., Tam, J., Yang, M. *et al.*, Technospheric mining of rare earth elements from bauxite residue (red mud): Process optimization, kinetic investigation, and microwave pretreatment. *Sci. Rep.*, 7, 15252, 2017.
60. Wahg, A.S. and Pinnock, W.R., Occurrence of scandium and rare earth elements in Jamaica Bauxite waste. *Econ. Geol.*, 82, 757–761, 1987.
61. Logomerac, V.G., Distribution of rare earth and minor elements in some bauxide and red mud produced, in: *Proceedings of the second international symposium of ICSOBA*, Budapest, Hungary, October 6-10, 1971, vol. 3, pp. 383–939, 1971.
62. Badwal, S.P.S., Ciacchi, F.T., Milosevic, D., Scandia-zirconia electrolytes for intermediate temperature solid oxide fuel cell operation. *Solid State Ionics*, 136, 91–99, 2000.
63. Petrakova, O., Klimentenok, G., Panov, A., Gorbachev, S., Application of modern methods for red mud processing to produce rare earth elements. *Proceedings of the 1st European Rare Earth Resources Conference (ERES 2014)*, Milos (Greece), September 4-7, 2014, pp. 221–229, 2014.
64. G.D. Fulford, G. Lever, T. Sato, Recovery of rare earth elements from red mud. US Patent 5,030,424, 1991.

    Gambogi, J., U.S. geological survey, 20202020. https://pubs.usgs.gov/ periodicals/mcs2020/mcs2020-rare-earths.pdf. [Accessed, July 18, 2022].
65. Ochsenkühn-Petropoulou, M., Lyberopulu, T., Ochsenkuhn, K.M., Parissakis, G., Recovery of lanthanides and yttrium from red mud by selective leaching. *Anal. Chim. Acta*, 319, 249–254, 1996.
66. Ochsenkühn, M.T., Hatzilyberis, K.S., Mendrinos, L.N., Salamas, C.E., Pilot plant investigation of the leaching procedure for recovery of scandium from red mud. *Ind. Eng. Chem. Res.*, 41, 5794–5801, 2002.
67. R. Boudreault, J. Fournier, D. Primeau, Processes for treating fly ashes. Patent WO 2013142975, 2013a.

68. Pradip, K. and Fuerstenau, D.W., Alkyl hydroxamates as collectors for the flotation of bastnasite rare earth ores, in: *Rare Earths*, R.G. Bautista, and M.M. Wong, (Eds.), pp. 57–70, The Minerals, Metals and Materials Society, Warrendale, PA, 1988.
69. Preston, J.S., The recovery of rare earth oxides from a phosphoric acid byproduct. 4. The preparation of magnet-grade neodymium oxide from the light rare earth fraction. *Hydrometallurgy*, 42, 151–167, 1996.
70. Lokshin, E., Vershkova, Y.A., Vershkov, A., Tareeva, O., Leaching of lanthanides from phosphohemihydrate with nitric acid. *Russ. J. Appl. Chem.*, 75, 1753–1759, 2002.
71. Habashi, F., The Recovery of the Lanthanides from Phosphate Rock. *J. Chem. Technol. Biotechnol. A*, 35, 5–14, 1985.
72. Habashi, F., Pyrometallurgy, in: *Principles of Extractive Metallurgy*, vol. 3, Gorden and Breach, New York, 1986.
73. Uysal, E., Al, S., Emil-Kaya, E., Stopic, S., Gürmen, S., Friedrich, B., Hydrometallurgical recycling of waste NdFeB magnets: Design of experiment, optimization of low concentrations of sulphuric acid leaching and process analysis. *Can. Metall. Q.*, 2022.
74. Lee, C.H., Chen, Y.J., Liao, C.H., Selective leaching process for neodymium recovery from scrap Nd-Fe-B magnet. *Metall. Mater. Trans. A Phys. Metall. Mater. Sci.*, 44, 13, 5825–5833, 2013.
75. Lyman, J.W. and Palmer, G.R., Recycling of rare earths and iron from NdFeB magnet scrap. *High Temp. Mater. Process.*, 11, 1–4, 175–188, 1993.
76. Tian, Y., Liu, Z., Zhang, G., Recovering REEs from NdFeB wastes with high purity and efficiency by leaching and selective precipitation process with modified agents. *J. Rare Earths*, 37, 2, 205–210, 2019.
77. Fomin, E.S., Sabirzyanov, N.A., Anashkin, V.S., Yatsenko, S.P., Diev, V.N., New approaches to processing of the muds of alumina production. *Travaux ICSOBA*, vol. 31, pp. 66–71, 2004.
78. Jiang, L., Cai, C., Worden, R.H., Li, K., Xiang, L., Chu, X., Shen, A., Li, W., Rare earth element and yttrium (REY) geochemistry in carbonate reservoirs during deep burial diagenesis: Implications for REY mobility during thermochemical sulfate reduction. *Chem. Geol.*, 415, 87–101, 2015.
79. Gupta, C.K., Pyrometallurgy, in: *Chemical Metallurgy: Principles and Practice*, C.K. Gupta, (Ed.), Wiley-VCH, 2003.
80. Whitworth, A.J., Vaughan, J., Southam, G., van der Ent, A., Nkrumah, P.N., Ma, X., Parbhakar-Fox, A., Review on metal extraction technologies suitable for critical metal recovery from mining and processing wastes. *Miner. Eng.*, 182, 107537, 2022.
81. Fathollahzadeh, H., Eksteen, J.J., Kaksonen, A.H., Watkin, E.L.J., Role of microorganisms in bioleaching of rare earth elements from primary and secondary resources. *Appl. Microbiol. Biotechnol.*, 103, 1043–1057, 2019.

82. Vaughan, J., Riggio, J., Chen, J., Peng, H., Harris, H.H., van der Ent, A., Characterisation and hydrometallurgical processing of nickel from tropical agromined bio-ore. *Hydrometallurgy*, 169, 346–355, 2017.
83. Gupta, C.K. and Krishnamurthy, N., Extractive metallurgy of rare earths. *Int. Mater. Rev.*, 37, 5, 197–248, 1992.
84. Demol, J., Ho, E., Soldenhoff, K., Senanayake, G., The sulfuric acid bake and leach route for processing of rare earth ores and concentrates: A review. *Hydrometallurgy*, 188, 123, 2019.
85. Davris, P., Balomenos, E., Panias, D., Paspaliaris, I., Leaching of rare earths from bauxite residues using imidazolium based ionic liquids. *Proceedings of the 1st European Rare Earth Resources Conference (ERES 2014)*, Milos (Greece, September 4-7, 2014, pp. 241–252, 2014.
86. Chaney, R.L., Baker, A.J., Morel, J.L., The long road to developing agromining/ phytomining, in: *Agromining: Farming for Metals*, pp. 1–17, Springer, Berlin, 2018.
87. Nkrumah, P.N., Echevarria., D., Erskine, P.G., Van der, A., *Extracting Innovations: Mining, Energy, and Technological Change in the Digital Age*, M.J. Clifford (Eds.), pp. 313–324, CRC, Boca Raton, 2018.
88. Van der Ent, A., Parbhakar-Fox, A., Erskine, P.D., Treasure from trash: Mining critical metals from waste and unconventional sources. *Sci. Total Environ.*, 758, 143673, 2021.
89. Van der Ent, A., Baker, A.J.M., Reeves, R.D., Pollard, A.J., Schat, H., Hyperaccumulators of metal and metalloid trace elements: Facts and fiction. *Plant Soil*, 362, 1, 319–334, 2013.
90. Liu, J.L., Yin, Z.L., Li, X.H., Hu, Q.Y., Liu, W., Recovery of valuable metals from lepidolite by atmosphere leaching and kinetics on dissolution of lithium. *Trans. Nonferrous Met. Soc. China*, 29, 3, 641–649, 2018.
91. Chour, Z., Laubie, B., Morel, J.L., Tang, Y., Qiu, R., Simonnot, M.O., Muhr, L., Recovery of rare earth elements from dicranopteris linearis by an enhanced ion 545 exchange leaching process. *Chem. Eng. Process.*, 130, 208, 2018.
92. Abhilash, S.S., Sinha, M.K., Pandey, B.D., Extraction of lanthanum and cerium from Indian red mud. *Int. J. Miner. Process.*, 127, 70–73, 2014.
93. Bani, A., Echevarria, G., Sulçe, S. *et al.*, *In-situ* phytoextraction of Ni by a native population of *Alyssum murale* on an ultramafic site (Albania). *Plant Soil*, 293, 79–89, 2007. https://doi.org/10.1007/s11104-007-9245-1.
94. Chaney, R.L., Angle, J.S., Broadhurst, C.L., Peters, C.A., Tappero, R.V., Sparks, D.L.J., Improved understanding of hyperaccumulation yields commercial phytoextraction and phytomining technologies. *J. Environ. Qual.*, 36, 1429, 2007.
95. El-Didamony, H., Ali, M.M., Awwad, N.S., Fawzy, M.M., Attallah, M.F., Treatment of phosphogypsum waste using suitable organic extractants. *J. Radioanal. Nucl. Chem.*, 291, 907–914, 2012.

96. International Energy Agency, The role of critical minerals in clean energy transition, 2021. TheRoleofCriticalMineralsinCleanEnergyTransitions.pdf [Accessed July 15, 2022].
97. Vaculchuck, R., Isataeva, A., Kolondzinkaia, G., Overland, I., Sabyrbekov, R., *Fossil Fuels in Central Asia: Trends And Energy Transition Risks*, vol. 28, pp. 1–6, Central Asia Regional Data Review, 2022, http://dx.doi.org/10.13140/RG/2.2.11461.37607.
98. International Energy Agency, *African Energy Outlook: World Energy Outlook Special Report*, 2019, https//:www.iea.org/reports/africaenergyoutlook. [Accessed, July 27, 2022].
99. Merritt, R.R., High temperature methods for processing monazite: I reaction with calcium chloride and calcium carbonate. *J. Less Common Met.*, 166, 197–210, 1990.
100. Blissett, R.S., Smalley, N., Rowson, N.A., An investigation into six coal fly ashes from the united kingdom and poland to evaluate rare earth elements content. *Fuels*, 119, 236–239, 2013.
101. Jha, M.K., Kumari, A., Panda, R., Kumar, J.R., Yoo, K., Lee, J.Y., Review on hydrometallurgical recovery of rare earth elements. *Hydrometallurgy*, 6, 1, 2–26, 2016.
102. Bleiwas, D.I. and Gambogi, J., *Preliminary Estimates of the Quantities of Rare-Earth Elements Contained in Selected Products and in Imports of Semi Manufactured Products to the United States, 2010*, pp. U.S. Geological Survey Open–File Report 2013–1072, 14, 2013, http://pubs.usgs.gov/of/2013/1072.

Index

Also of Interest

Check out these other books by Elvis Fosso-Kankeu published by Scrivener Publishing

Synergy of Bio-Chemical Processes for Photocatalytic and Photoelectrochemical Wastewater Treatment
Edited by Sadanand Pandey, Elvis Fosso-Kankeu and Soumya Pandit
Forthcoming 2024. ISBN 9781394197873

Recovery of Values from Low-Grade and Complex Minerals Development of Sustainable Processes
Edited by Elvis Fosso-Kankeu, Bhekie B. Mamba and Antoine F. Mulaba-Bafubiandi
Published 2024. ISBN 9781119896418

Hybridized Technologies for the Treatment of Mining Effluents
Edited by Elvis Fosso-Kankeu and Bhekie Mamba
Published 2023. ISBN 978-1-119-89642-5

Photoreactors in Advanced Oxidation Processes The Future of Wastewater Treatment
Edited by Elvis Fosso-Kankeu, Sadanand Pandey, Suprakas Sinha Ray
Published 2023. ISBN 978-1-394-16629-9

Photocatalysts in Advanced Oxidation Processes for Wastewater Treatment
Elvis Fosso-Kankeu, Sadanand Pandey and Suprakas Sinha Ray
Published 2020. ISBN 978-1-119-63139-2

Recovery of Byproducts from Acid Mine Drainage Treatment
Edited by Elvis Fosso-Kankeu, Christian Wolkersdorfer and Jo Burgess
Published 2020. ISBN 978-1-119-62007-5

Nano and Bio-Based Technologies for Wastewater Treatment
Prediction and Control Tools for the Dispersion of Pollutants in the Environment
Edited by Elvis Fosso-Kankeu
Published 2019. ISBN 978-1-119-57709-6

www.scrivenerpublishing.com

Printed in the USA/Agawam, MA
September 16, 2026

910661.062